Water's Edge

Like water, the contributions to
this book take various forms.

They draw from science, memoir, trance, poetry, illustration, and
years of research. One essay channels the voice of a salamander.
Other works reference South Africa, Australia, Brazil, Mexico,
Dominican Republic, Namibia, India, Spain, the diverse
geographies in the United States.

Rather than the authors concentrating and
pursuing a singular, overwhelming argument,
they circulate moments of apprehension,
intimation, and felt experience.

They are like tributaries: each essay, poem or
image carries themes of water in a distinctive
style, offering exigent and often intimate
reports on the substance upon which all living
organisms depend—water.

≈

Water's Edge

Writing on Water

EDITED BY

LENORE MANDERSON

AND

FORREST GANDER

CURBSTONE BOOKS/NORTHWESTERN UNIVERSITY PRESS
EVANSTON, ILLINOIS

Curbstone Books
Northwestern University Press
www.nupress.northwestern.edu

The editors are grateful to the Institute at Brown for Environment and Society (IBES) for core
funding and to the Marshall Woods Lectureship Foundation of Fine Arts at Brown University
for support for the colloquium "Writing on Water." For details of "Water's Edge" in 2018, see
lenoremanderson.com/waters-edge.

Coral Bracho, "Water's Lubricious Edges," translated by Forrest Gander, from *Firefly under the
Tongue: Selected Poems of Coral Bracho*, copyright © 2008 by Coral Bracho, translation copyright ©
2008 by Forrest Gander. Reprinted by permission of New Directions Publishing Corp.

Colin Channer, "Roots," from *Liberties Journal of Culture and Politics* 2, no. 2 (Winter 2022), edited by
Leon Weiseltier, copyright © 2022. Reprinted by permission of the Liberties Journal Foundation.

Forrest Gander, "Sangam Acoustics: Immigrant Sea," from *Twice Alive*, copyright © 2019, 2020,
2021 by Forrest Gander. Reprinted by permission of New Directions Publishing Corp.

Maya Khosla, "Giant Tuna," from *Keel Bone*, copyright © 2003 by Maya Khosla. Reprinted by
permission of the author.

Elizabeth Rush, "Atlas with Shifting Edges," from *Emergence Magazine*, June 25, 2019. Copyright
© 2022 by *Emergence Magazine*, an initiative of Kalliopeia Foundation. All rights reserved.

Printed in the United States of America

10   9   8   7   6   5   4   3   2   1

Library of Congress Cataloging-in-Publication Data

Names: Manderson, Lenore, editor. | Gander, Forrest, 1956– editor.
Title: Water's edge : writing on water / edited by Lenore Manderson and Forrest Gander.
Description: Evanston, Illinois : Curbstone Books/Northwestern University Press, 2023. | In
English; 2 contributions in original Spanish and English translation. Provided by publisher.
Identifiers: LCCN 2022023699 | ISBN 9780810145795 (paperback) | ISBN 9780810145801 (ebook)
Subjects: LCSH: Water—Poetry. | Water—Fiction. | Water—Social aspects.
Classification: LCC PN6071.W37 W38 2023 | DDC 808.8036—dc23/eng/20220524
LC record available at https://lccn.loc.gov/2022023699

Contents

Preface

LENORE MANDERSON AND
FORREST GANDER

WE LIVE ON A PLANET OF WATER. WHAT WE CALL "LIFE"—OUR LIVES AND the lives of everything on Earth—is dependent on it. The average adult human body is 60 percent water, while infants are closer to 75 percent. Although by mass, Earth is not watery, about 71 percent of its surface is covered by water; of that, only about 3 percent is freshwater. Most is locked into glaciers, floating in the atmosphere, or trapped in soil or underground. Yet in one form or another, water is quotidian, ubiquitous, precious, and precarious.

Heraclitus is a conventional reference point: "You cannot step into the same river twice," or alternatively, "We step and do not step into the same river—we are and we are not." His reference is to water's elusiveness and illusion, to time, place, and space. Water is always changing, and we in relation to it. We find meaning in our varied encounters with its many forms—the smell of fresh rain, the clatter of hail—its containment and formlessness. Water moves in ebbs and flows, rips and tides, waves and currents. The expansiveness of oceans contrasts with the accessibility of a small pond, a major river with a small stream. Water is an element of contrasts. It is rarely blue, or rather, it only sometimes appears blue: it is also green, yellow, brown, white, gray, indigo, and black, and in the end, clear—a colorless element that takes its hue from the environment of which it is part.

Water draws us across a wide environment as rivers trace terrain, territory, and state; oceans stake their edges. But water runs its own course. While engineering has sought to harness water for human purposes, for agriculture, domestic and industrial use, and for electricity, still, whether or not there is water, and how much there is, is largely beyond our control.

In writing this, we are mindful of the destructiveness of droughts in our home geographies—Forrest on the West Coast of the United States, Lenore in Australia and South Africa. The absence of water reminds us of its scarcity and unpredictability. In California, destructive wildfires raged even as we collaborated on this volume. Australia's droughts destroyed rich grazing land and kindled fires that engulfed bushland, farms, and towns. South Africa's droughts continue to draw sharp edges around the inequalities of everyday life—for the past three years, in Cape Town; now, as we write, other cities and towns are in a state of emergency, the minimum for everyday survival trucked in and bought and sold. Where there is no water, how can people handwash as enjoined in times of pandemic? How can they nourish crops? How far must they walk to find potable water? We learn, in points across the globe, not to waste what we cannot live without.

~~~

Water's Edge has its roots in meetings on the elements. From 2015, Lenore curated and produced a program called "Earth, Itself" for the Institute at Brown University on Environment and Society (IBES). "Earth, Itself" was intended to highlight scholarly and artistic contributions to ways of understanding earth systems and elements, environmental change, knowledge systems, and policy; through interdisciplinary exchanges, we aimed to address some of the most complex problems of our time in innovative, engaging, and inclusive ways. The gatherings and exhibitions were designed to facilitate conversations and nurture debate and collaboration across the creative arts and humanities, the social and natural sciences.

with a very particular choice of vocabulary and phrasing. Language adds tempo and rhythm, and thus affect. We come to feel the water, and the environments of which water is part, through text. To write about water is to write about how it feels and tastes, how it looks and sounds. The language of water is as visceral as the element, the words written to be rolled around the mouth. The prosody of the contributions in this volume plays with the boundaries by which we differentiate literary forms.

These works capture our original goals to explore and celebrate water, while reflecting on its disturbances: the pollution of riverways; the drying out of earth and the depletion of food and potable water in some places; in other places, the destructiveness of storms, rapid floods, and the slow encroachment of saltwater on land that destroys human, plant, and animal habitats.

Water is everywhere, and yet visible only through its impurities, refractions of light, and reflections of solid images. Water is fundamental to all life forms, and life on Earth is vulnerable precisely because of threats to predictability and renewability, quantity and quality. Here, writers from different genres invite us to dive into their work, as they consider water's plentitude and paucity, its elementalism and allusiveness, and our relations to it in its many forms.

Water's Edge

Sangam Acoustics: Immigrant Sea

FORREST GANDER

Aroused by her inaccessibility, he aches for more
of her life to live inside him. Watching

the breakers, standing so close he can
feel heat coming off her wet scalp. What is

his relation to this person
before him, so familiar and foreign? The way

he searches out her face, he searches out himself. Gusts
thrash crests of swell, spring grasses twirl

circles in the sand where they stand without speaking. She
wants him to know it's all charged, even grass

positive, pollen negative, so when grass waves,
it sweeps the air for pollen. He feels electricity all around

as though the wild drama of the coming storm were already
aware of them, foreigners on this shore. Little

sapphire-blue flowers speckle the dunes.
He wonders if he has let himself flatten out

into a depthless sheet, like escalator stairs, whether in the end
he'll disappear underground without the smallest lurch

of resistance. But when her lavish face turns toward him
beaming, the corners of her eyes wind-wet,

he yields to that excess, he reappears to himself.

The Lake and the City

José-Luis Moctezuma

> Far from the immensities of sea and land, merely through memory, we can recapture, by means of meditation, the resonances of this contemplation of grandeur. But is this really memory? Isn't imagination alone able to enlarge indefinitely the images of immensity?
>
> —Gaston Bachelard, *The Poetics of Space*

> Surely
> Upon the whole length of my form
> The water birds will alight
>
> —Anishinaabeg Dream Song, as sung by Gegwejiwebinan

The edge of Lake Michigan is ringed with the sediment of time. I walk along its shore and conjure the city that borders it. The lake works like a mirror, and my feet in their movement, walking ceremoniously along a sandbar as if I were walking down the nave of an open-air cathedral, misplace me in a certain slant of light. The city instead conjures me, and the lake on its side lies there, a body imperceptible in its entirety; its waters soundlessly stir in a shared remembrance of ancestral glaciers, like the leaky residue of an immense ice cube encountering the pangs of a primordial summer, or a scattered dream of elephants parading upon the shell of a world-spanning tortoise, the dream-fabric already partially torn upon the first minutes of awakening to another work-saturated day. The name of the city is immaterial to the facticity of the lake, whose supreme objecthood places a fiction at the center of this meditation, which is also a dream, a reconstruction of the memoranda of water. The city might be Milwaukee, it might be Waukegan, it might be Saugatuck, their tristate thread luminous at night in the satellite's gaze. The city might be invisible to the eye, but it is there in the mind in the way the lake is there in the mind, and the stars are imprinted on the surface of the lake at midnight's passage, in the place where the Michigan is as dark-hearted as a speck of gold inside a crumb of coal within the belly of a mineshaft.

In the local instance, the city is Chicago, where I live, but the name of the city, immaterial to the material supremacy of the lake, extends itself (we may say, fabricates itself) in the way the Chicago River extends the grid of Chicago, or in the way the lake extends past the mouth of the city. It is not only Chicago but other cities that flow into and out of each other along the lakeshore, in the sense that water flows into and out of the illusory immovability of housing structures.

The figure of the city and the figure of the lake are, in actuality, one hydrologic motion between elemental states that posit contradictory expressions of immobility and kinematics, of the silence of a reservoir and the clamor of flow. Paradoxically, these figures switch roles, depending on the time of day or season. In the inky depths of the night, the lake is alive, throbbing and lapping against the shore in an eros of space, the city asleep with the wind-down of the business of the day. During the day, the city bursts to life and the noise of the everyday streams and flows in electrical charges into and out of the streets and buildings; the lake, stolid and massive in the blinking gaze of the sun, shifts in meek currents, absorbing the magnetic fields of the city's waking life and dreaming of the moon's currents to come.

We call the network of cities interlaced along the shores of the Michigan, Superior, Huron, Erie, and Ontario the Great Lakes Megalopolis, and the Great Lakes birth these hubs of civitas, culture, and capital with the ease of a river breaking over stone and uncovering the concealed pigmentations and densities of the soil and what pressure-loosened dreams of the earth rise from hazards of geology. If the city is multiple, the lake too is multiple. On the side of Superior, Michigan, and Erie, the Great Lakes radiate out from their heart in the Straits of Mackinac—in Chippewa, *Michilimackinac*—circling the great turtle shell of an island that serves as the tip of a leviathan consciousness. The city of Chicago is a centrifuge, its heart in the center of Lake Michigan; Chicago opens its arms outward from the calyx of the lake's voluminous stem. Dwelling on the mirror image of water in concert with the moon, Brenda Hillman writes:

> & the word described
> as water came
> in the style of a stem
> by a stream
> by a stone[1]

Chicago rings Lake Michigan as a supernal mycelium in which multitudinous forms of life reimagine rituals of language in a gridwork of glass and voice, pavement and bodies, electrical signage and vanishing faces, the public intimacy of train cars passing each other and waving aloft in the night air. In Green Bay, in Gary, in Muskegon, similar phenomena unfold and evaporate, and the lake subtends them each as the northwest wind subtends the transmission and gradual depositing of sand dunes on the coast of Indiana, as the lake lowers its garment (the temperature of the earth lowering and raising the sea levels) and unveils new layers of striated beach ridge.

In John Cage's *A Dip in the Lake: Ten Quicksteps, Sixty-two Waltzes, and Fifty-six Marches for Chicago and Vicinity* (1978), the fungal colony of the city is alchemized into a surfeit of neatly rigid color lines drawn in felt pen on a map of Chicagoland. The lines overwhelm and saturate the grid of the city, blurring its cartography in a palimpsest of chance operations that Cage famously would employ as his compositional practice. The map is not the territory, but it serves as a musical score for a network of voices and sounds that weave together the city's holographic shape. In Cage's aleatoric composition, the lake and the city are a single, streamlined entity, a reciprocation of spatial tensions in which lines chaotically crisscross, as if it were possible to walk into and stand in the lake as one would on a city street. For Cage, a "dip in the lake" is possible because the map makes it possible: they are supreme fictions activated by a belief evidenced in the walking there, on the lake shore staring out, taking a dip in the lake with one's mind. Cage, of course, literally intended for these randomized coordinates to be visited by an enterprising team of players and recorders (as accomplished in Toronto, another Great Lake city). But the mental landscaping is noteworthy, recalling Wallace Stevens:

> Perhaps
> The truth depends on a walk around a lake,
>
> A composing as the body tires, a stop
> To see hepatica, a stop to watch
> A definition growing certain and
>
> A wait within that certainty, a rest
> In the swags of pine-trees bordering the lake.[2]

The lake constructs a border zone between the elemental nature of water and the elemental force of the city. The border zone is permeable and imaginary, yet it is there. We might regard the city, the megalopolis, as a fifth element, a calcification of time stretched along the pebble and silt of exposed lake bed, like the lacework of uprooted seaweed glistening in the sun's heavy-breathing gaze. The city, they say, is an artificial paradise, the "imposition on the world of structures it never asked for."[3] The lake is a natural body whose memory transcends the margins of place and placeme(a)nt. Or it may be that the city limns a natural phe-nomenon in which the lake is revealed to be pure artifice, by which I mean that the lake acts as a conjurer, a mind behind the animate veil of wavelengths, and its phantasmal figuration achieves the sensory qual-ities of the living (it is alive in a way we are not), when we are asleep, miles inland, and the silence of no wave lapping against an empty rock face reaches us from the abyss of a city's unconscious.

The edge of Lake Michigan is ringed with the concretion of space. I am not walking along its shore because it is the pandemic time, the time of quarantine. We shelter not in place but in memory. The memory is that of the color of lake water at 5:14 in the afternoon, gone before anyone realizes. The lake is still there, despite the fact that I am not near its water, nor walking its shore. The falling snow distorts my vision only a few meters past my window; the ice buries the lake in a white mosaic. Lake Michigan is always on the verge of becoming ocean. The spiritual bathymetry of a lake defies our human understanding. Bachelard, dis-cussing the work of Philippe Diolé and deep-sea diving, notes that "we are so remote from the earth and life on earth [when deep-sea diving], that this dimension of water bears the mark of limitlessness. To try to find high, low, right or left in a world that is so well unified by its sub-stance, is thinking, not living."[4] Seen this way, the lake can be reframed as the very substance, in its unbreakable solidity, of intellection and thought. It embodies a consistency of peacefulness that the ocean, in its emotional volubility and capricious temperament, does not quite have. The lake is a medium in the sense that vision is a medium. Where vision blurs into uncertainty, the lake is no longer water, but water's abstraction, and diving beyond the surface and cut of wave teaches one how to value breath as breath becomes impossible. One is sunk not in water but in the outer reaches of thought, where the breath collects in the lungs before bursting into song as one reaches the sur-face. In the time of pandemic, we've learned that breath and breathing matter most.

~

There's a Tlingit song, composed by Hayi-a'k!ᵘ, whose name translates to "Small-Lake-Underneath." The song—entitled "The Mourning Song of Small-Lake-Underneath"—goes:

> I always compare you to a drifting log with iron nails in it.
> Let my brother float in, in that way.
> Let him float ashore on a good sandy beach.
> I always compare you, my mother, to the sun passing behind the
>   clouds.
> That is what makes the world dark.[5]

The transportation system of the metaphor is used here literally in describing the transport of the poet's brother on a funereal log drifting out into the arms of the sea. It is echoed by the transport of their mother to the mobility of the sun passing behind clouds, producing a deepened pall of the emotions. The element of water backgrounds such vehicularity, and the ocean-sea-lake metaphor of metaphors remains invisible in Hayi-a'k!'s song because it is the already-given of metaphoricity, the very beating heart of transference between death and life.

In *The Jesuit Relations*, the annual chronicles of Jesuit missions and misadventures in "New France" during the colonial moment of the seventeenth century, we encounter a similar metaphoricity, this time involving Jean Nicolet living among the Nipissings and the Hurons, and translating French and the Algonquian languages to missionaries and merchants. Nicolet and the missionaries spend their days catechizing the Indigenous in the doctrinal folklore of Christianity; the waters of the Great Lakes, threaded like beads on a rosary by the St. Lawrence River, likely served as a medium for baptismal rituals, for the heterogeneous economies of the fur trade, and for the unwanted arrival of bearded white men and wilderness debutants. In response to a Jesuit priest's description of the eclipse of the sun during Christ's crucifixion, Huron "seminarists" describe a similar darkening of the earth:

> "There is still talk" (said they) "of a very remarkable darkening of the Sun, which was supposed to have happened because the great turtle which upholds the earth, in changing its position or place, brought its shell before the Sun, and thus deprived the world of sight."[6]

The limbs of the cosmic tortoise, it may be said, can be traced in the radial overflow of the Great Lakes Basin. Our first "Lake Poets," who envisioned the tortoise in the contours and isobaths of the lake, were the Anishinaabe who lived on its shores. The First Nations Lake Poets did not see merely the pathos and fragmentation of human history passed and passing, but the eternal present of the Great Water and its tributary force in the oral culture of the people. Ojibwe poet Jane Johnston Schoolcraft (her Ojibwe name, Bamewawagezhikaquay) wrote the following along the banks of Lake Superior:

> Here in my native inland sea
> From pain and sickness would I flee
> And from its shores and island bright
> Gather a store of sweet delight.
> Lone island of the saltless sea!
> How wide, how sweet, how fresh and free
> How all transporting—is the view
> Of rocks and skies and waters blue
> Uniting, as a song's sweet strains
> To tell, here only nature reigns.[7]

Schoolcraft wrote her verses in both Ojibwe and English, and often translated her Ojibwe work into English, and although the original Ojibwe version of that particular poem did not survive, its lines appear scripted on the fabric of lake water, not as the tragedy of Romantic identity, but as the protean grandiosity of the lake's permanence. Her husband, Henry Rowe Schoolcraft, a geographer and ethnologist who published texts on the Indigenous cultures of the Great Lakes region, contributed with her to the source material for Longfellow's epic *The Song of Hiawatha*. The writings of Schoolcraft's husband would also form part of the source material for the lyric work of another Great Lake Poet, Lorine Niedecker, who generations later would compose *Lake Superior*, a long poem whose compression of form disguises the tremendous amount of research, citation, and intertextuality that fed into its thirteen-part structure. Looking out from the shores of Wisconsin, Niedecker describes the materiality of the lake in the figuration of the stone:

> In every part of every living thing
> is stuff that once was rock
>
> In blood the minerals
> of the rock[8]

One can read here the absence of the lake in bridging and piecing together time's fragments, documents of colonial history and colonial violence. Niedecker meditates on the former wholeness of what had once the hardness of rock, signaling the precedence of water, of the lake's unstoppable alchemy, to break down, fragment, wreck and drown, and (in a sense) rebirth. This is history's memoranda within the geological circuitry of the land's formation, through a *via negativa*, in its gradual effacement and ridgework by the lake. It is astonishing that the lake is what remains (in)visible in *Lake Superior*, as water remains invisible to the perch, trout, and bass that swim in its waters—it is the backdrop to which Niedecker reconstructs a history of the (colonialist) enclosure movement and its resulting nature/culture divide. The Schoolcrafts make an appearance in *Lake Superior*, and what the geographer misses, the Ojibwe Lake Poet catches:

> I'm sorry to have missed
> Sand Lake
> My dear one tells me
> we did not
> We watched a gopher there[9]

But the lake, as a force and as a presence, might also be (seemingly) invisible because Niedecker already places the mnemonic powers of freshwater in the medium of poetic vision. In another poem, Niedecker remarks on the watery qualities of human memory (and the body after all is 60 percent composed of water), as mimicked in the desire which the lake manifests in the vision of the Lake Poet peering over the artificial (illusory) border the city erects between itself and the lake:

> We are what the seas
> have made us
>
> longingly immense
>
> the very veery
> on the fence[10]

Daniel Borzutzky's *Lake Michigan* contains a lake within a lake, in an epigraph by Simone de Beauvoir, in which de Beauvoir witnesses the city of Chicago alchemized into a wondrous lake scene that brings to mind "the luxury of the Cote d'Azur." For a moment, there was "nothing to remind her of the squalor with its human wreckage" that Chicago in that historical instance represented.[11] If Lake Michigan signals an escape from the sociopolitical reality of Chicago for de Beauvoir, Borzutzky's *Lake Michigan*, pierced through with scenes of racial and economic segregation, "black sites" of city-funded torture, and the unpunished killing of Black and Brown bodies, brims over with human wreckage. The lake can be experienced completely differently depending on which part of the shore you look out from. We thus witness two alternate realities, two distinct cities and two distinct lakes, in an optical illusion generated by an imbrication of competing ecologies. The city is a compressed network predicated on the lake's ecological precedence, and Borzutzky draws our attention to the mute witness of Lake Michigan on the horrors of the city's policing of bodies deprived of name, rights, and citizenship. The lake's metatextual indifference to human affairs reflects the intratextual indifference of the police and carceral state on human dignity. What Poe earlier referred to as "the terror of the lone lake," Borzutzky updates to the level of the grotesque:

> It is raining again on Lake Michigan
> Some say it is raining bodies but really it is trash[12]

Base and superstructure in the relationality of the lake to the city, and the city to the lake, reverse and switch depending on standpoint. Human wreckage inspires nonhuman spillage into the lake's body. The neoliberal mandates of global capitalism, whether in the grand abstraction of the Great Lakes Basin or in a country like Nicaragua, its growth crippled by decades of U.S. American imperialism and proxy dictatorships, funnel the same refuse of the megalopolis, a network of cities, into the reciprocal network of rivers, lakes, and oceans. Ernesto Cardenal, in "New Ecology," citing Somoza's US-backed dictatorship, describes the devastation of the two ecosystems, civic and natural, in the aftermath of "Somocism":

José Somoza was wiping out the sawfish of the Great Lake.

. . .

And poor Rio Chiquito! Its misfortune
the whole country's. Somocism mirrored in its waters.
The Rio Chiquito in Leon, fed by streams
of sewage, wastes from soap factories and tanneries,
white water from soap factories, and red from tanneries;
plastics on the bottom, chamber pots, rusty iron. Somocism
left us that.

. . .

And into Lake Managua all of Managua's sewage water
and chemical wastes.[13]

In Chicago, the ingenuity of the architects and engineers has, at least, spared Lake Michigan too grotesque a backflow of civic waste. Cardenal's injunction, whether with Lake Managua or Lake Michigan in mind, is that the border between the lake and the city, between the local and the global, must recognize the rights of all human and nonhuman life that intersect at the crossroads of the Great Lakes Megalopolis:

We're going to decontaminate Lake Managua.
The humans weren't the only ones who longed for liberation.
The whole ecology had been moaning. The Revolution
also belongs to lakes, rivers, trees, animals.[14]

In this vision of the "lake ethic," we might see a renewed interplay of the intersectional qualities of LakeXCity life in a way that redefines our sense of the geopolitical: what the lake experiences will eventually be reflected in the life of the cities it feeds, fashions, and embraces. Following the words of another Anishinaabe–White Earth Nation poet, Kimberly M. Blaeser, we must go down into the water to understand what the water is:

And sometimes in our water dreams
we pitiful land-dwellers
in longing
recall, and singing
make spirits ready
to follow:
bakobii.[15]

NOTES

Epigraphs: Gaston Bachelard, *The Poetics of Space*, trans. Maria Jolas (Boston: Beacon Press,1994), 183. Annishinaabeg dream song, as translated by Mary Warren English. In the literal translation by Margaret Noodin, the song is transcribed as one sentence without line breaks: "It is certain they land on me the thunderbirds across my existence." Both translations in *When the Light of the World Was Subdued, Our Songs Came Through: A Norton Anthology of Native Nations Poetry*, edited by Joy Harjo (New York: W. W. Norton & Co., 2020), 18.

1. Brenda Hillman, "November Moon," *Practical Water* (Middletown, Connecticut: Wesleyan University Press, 2009), 59.

2. Wallace Stevens, "Notes Toward a Supreme Fictione," *Collected Poems* (New York: Vintage Books, 1990), 386.

3. Rem Koolhaas, *Delirious New York* (New York: Monacelli Press, 1994), 246.

4. Bachelard, *Poetics of Space*, 205.

5. Jerome Rothenberg, ed., *Technicians of the Sacred: A Range of Poetries from Africa, America, Asia, Europe, and Oceania* (Oakland: University of California Press, 2017), 94.

6. Reuben Gold Thwaites, *Jesuit Relations*, Vol. XII (Cleveland: The Burrows Brothers Company, 1637), 71–72.

7. Jane Johnston Schooolcraft, "Lines Written at Castle Island, Lake Superior," in *The Sound the Stars Make Rushing through the Sky: The Writings of Jane Johnston Schoolcraft*, edited by Robert Dale Parker (Philadelphia: University of Pennsylvania Press, 2007), 92, lines 1-10.

8. Lorine Niedecker, *Lake Superior*, in *Collected Works*, ed. Jenny Penberthy (Berkeley: University of California Press, 2002), 232.

9. Niedecker, *Lake Superior*, in *Collected Works*, 237.

10. Niedecker, *Lake Superior*, in *Collected Works*, 240.

11. Borzutzky notes that the Simone de Beauvoir epigraph is taken from Lois Wille's *Forever Open, Clear, and Free: The Struggle for Chicago's Lakefront*.

12. Daniel Borzutzky, *Lake Michigan* (Pittsburgh: University of Pittsburgh Press, 2018), 73.

13. Ernesto Cardenal, "New Ecology," *Pluriverse*, ed. Jonathan Cohen (New York: New Directions Publishing, 2009), 176–77.

14. Cardenal, "New Ecology," in *Pluriverse*, 177.

15. The word *"bakobii"* means "Go down into the water." Kimberly M. Blaeser, "Dreams of Water Bodies" ["Nibii-Wiiyawan Bawaadanan"], translated by Margaret Noodin. In *When the Light of the World Was Subdued*, 67.

Chicken Shit and the Chesapeake Bay

ASHLEY DAWSON

I SPENT MY TEENS SURROUNDED BY CHICKEN SHIT, THOUGH MOSTLY I was unaware of this. Perdue and other agribusiness corporations made the Eastern Shore of Maryland the world headquarters for industrial-scale chicken manufacturing, but when my family moved there in the 1970s, Perdue kept their barns of birds carefully hidden on tracts of sequestered farmland. Now and again, my family would pass an eighteen-wheeler on the highway, stacked with thousands of small crates stuffed with miserable-looking chickens, their feathers flying in the wind as they barreled toward their doom. Perdue Farms grew into a titan of US agribusiness, and cheap chicken kept stomachs full, but the industrial chicken industry was largely invisible, out of sight and out of mind.

Nonetheless the chickens and their shit seeped into my life in ineluctable and ominous ways. After my family moved to the Eastern Shore from our home in South Africa, we bought a jaunty twenty-foot sailboat. Every weekend the family packed a picnic lunch and drove the sailboat on its trailer to some tributary river feeding into the mighty Chesapeake Bay, America's largest estuary, where we would maneuver the boat gingerly into the water and set off for a day's sailing. Looking back, these were magical days, my father's effort to re-create an Arthur Ransome–style idyll for his family in this unfamiliar new land. But at the time I felt nothing but deep dread. The possibilities for mayhem seemed endless; every moment spent on the boat was filled with fear of the horrible incidents likely to come. It wasn't as if my fears were based on unfounded concerns, like capture by pirates. No, my sense of dread was stoked by the much more concrete disasters that did befall us: being becalmed on the Sinepuxent Bay as the wind dropped in the afternoon, our small outboard motor sputtering short of gas, leaving us drifting helplessly into the mosquito-infested backswamps of Assateague Island; or capsizing as a ferocious squall blew up on the Nanticoke River, taking us unawares, throwing our possessions into the dark waters and trapping us under the waterlogged sail, gasping for air.

Our most memorable maritime disaster occurred not in the sailboat, but during a lull. We had sailed into the stretch of water just west of Roaring Point, where we were separated from the Chesapeake Bay by Bloodsworth Island. These evocative place names should have warned us of impending catastrophe, but my family, true to their settler colonial roots, sailed on blithely. Around lunchtime, we came on a small island, a kind of sandbar unmarked on any nautical map. My parents decided this was a good time to take a break; we lowered the sail and cast our small anchor overboard. My sister and I dove into the cool water and swam to the island, about thirty feet from the boat. Standing on the white sand, we waved jubilantly to our parents, and then dove back into the water to race back to the boat for lunch. As we swam back, we began hollering at our parents. They waved back again. But we were not waving happily—we had swum into a swarm of jellyfish, and our screams were of confusion and pain. By the time we made it back onto the boat, we were covered with red welts from the stings, venomous prints of jellyfish tentacles on our legs, our torsos, even our faces.

〰

While this one incident was not necessarily indicative of a shift in the health of the Bay, the heightened incidence of jellyfish in coastal waters is often a sign of eutrophication. When nutrients are added to coastal waters, massive algae blooms often take place. As they decompose, these algae soak up oxygen in the water, making it hard for fish to survive. But jellyfish do well in such anoxic conditions. Excessive nutrients in coastal waters also cause explosions of phytoplankton and zooplankton, the favorite food of jellyfish. Where do the excess nutrients that cause eutrophication come from? There are a variety of possible sources, including fertilizers applied to suburban lawns and agricultural cropland, but one of the major factors is waste. On the Eastern Shore, waste means chicken shit.[1]

During the 1980s and 1990s, more and more farmers, faced with declining income from corn, soybeans, and other crops, began working as subcontractors for Perdue, Tyson, and the other agribusiness giants that were coming to dominate the chicken industry in the US at the time. Today, Americans eat three times more chickens than they did fifty years ago.[2] In 2017, the Delmarva Peninsula, which abuts the Chesapeake, produced more than six hundred million chickens.[3] In order to dispose of the mountains of chicken waste generated as they built ever-larger chicken houses, farmers began to spray waste on their fields, waste that was eventually washed off the land and into freshwater tributaries of the Chesapeake Bay, like the Nanticoke and Wicomico rivers. We tended to avert our eyes and hold our noses as we passed the chicken death trucks on the highway and the giant chicken houses overrunning the countryside, but when we slid into the cool waters of the Chesapeake, we were literally swimming in chicken shit.

≋

The Chesapeake Bay is not the only place where eutrophication is an issue. The Stockholm Center for Resilience has identified nine planetary boundaries, the crossing of which will endanger life on our planet.[4] Three of these planetary boundaries should be seen as tipping points:

climate change, ocean acidification, and stratospheric ozone depletion. The other four environmental boundaries—the nitrogen and phosphorus cycles, freshwater use, changes in land use, and biodiversity loss—can be viewed as signifying the beginning of irreversible environmental degradation. We are at red alert status: three of these processes—climate change, the nitrogen cycle, and biodiversity loss—have already crossed their boundaries.

Of these three, the nitrogen cycle is the least well known, and yet it is absolutely fundamental to human life. The origin of the problem with nitrogen dates back to the years right before the First World War. Prior to this, the world's leading powers had engaged in increasing inter-imperial rivalry as they sought to control the international fertilizer trade, which had become essential to agriculture in the world's wealthy nations. In the early years of the century, researchers in Germany developed a process for fixing nitrogen from the atmosphere to produce ammonia, a key ingredient in artificial fertilizers for agriculture. Known after its inventors as the Haber-Bosch process, it helped industrialize agriculture on a massive, global scale. It also generated serious ecological problems. Today, 121 million tons of nitrogen are released into the environment each year, accumulating in waterways and helping to create expanding low-oxygen sites in coastal waters.

Known as dead zones, oceanic low-oxygen sites have quadrupled in size since 1950, while coastal dead zones have multiplied tenfold.[5] The Chesapeake Bay was one of the first sites where a significant marine dead zone was documented; today there are more than five hundred such sites around the world. Climate change is aggravating this trend toward deoxygenation of the oceans, since warmer waters hold less oxygen. The implications are stark: in the first comprehensive analysis of dead zones, a group of scientists wrote, "Major extinction events in Earth's history have been associated with warm climates and oxygen-deficient oceans. Under the current trajectory, that is where we would be headed."[6] The scientists conclude their report with a rather naive assertion that the consequences to humans of staying on this trajectory are so dire that "it is hard to imagine we would go quite that far down that path."[7]

The situation of the Chesapeake Bay is an important litmus test of this hope for the redemption of our coastal waters, and of life on Earth in general. Located right next to Washington, DC, the Chesapeake Bay is the most monitored body of water in the United States, its imperiled state a dramatic indication of the failure of political elites to attend to issues of environmental sustainability. Efforts to save the bay have achieved some dramatic successes over the last decade: in 2009, President Obama issued an executive order calling the bay "a national treasure" and ordering development of a federal plan to restore it; the following year the Environmental Protection Agency issued regulations limiting the amounts of nitrogen, phosphorus, and sediment that can enter the bay and its tidal waters; and in 2014 representatives from the entire watershed signed the Chesapeake Bay Watershed Agreement.[8] The American Farm Bureau Federation and other agricultural and real estate development organizations filed lawsuits to halt these plans to save the bay, but these suits were unsuccessful and dead zones in the bay began to shrink significantly, while underwater grasses that provide habitat to fish, crabs, and other marine species began to recover. But states like Pennsylvania are still sending tens of millions of pounds more nitrogen downstream than they have agreed to each year. Now, as industrial-scale chicken farms—each housing nearly forty thousand birds—overrun the Eastern Shore, the political power of Big Poultry is threatening the hard-fought environmental victories of the last decade.[9] With the Trump administration backing a massive wave of environmental deregulation and political corruption, my childhood memories of chicken-stuffed death trucks and jellyfish-filled waters take on fresh resonance and urgency.

NOTES

1. John Upton, "Can We Save Chesapeake Bay from Chicken Crap?," *Grist*, February 5, 2014. https://grist.org/food/can-we-save-chesapeake-bay-from-being-ruined-by-chicken-crap/

2. Georgina Gustin, "Giant Chicken Farms Overrun Delmarva, and Neighbors Fear It's Making Them Sick," *Inside Climate News*, April 23, 2018. https://insideclimatenews.org/news/23042018/poultry-industry-epa-asthmarespiratory-illness-chicken-houses-delmarva-legislature-health-study-delaware-maryland-virginia

3. Gustin.

4. John Bellamy Foster, Brett Clark, and Richard York, *The Ecological Rift: Capitalism's War on the Earth* (New York: Monthly Review Press, 2010), 14.

5. Damian Carrington, "Oceans Suffocating as Huge Dead Zones Quadrupled Since 1950, Scientists Warn," *The Guardian*, January 4, 2018.

6. Denise Breitburg et al., "Declining Oxygen in the Global Ocean and Coastal Waters," *Science* 359, no. 6371 (January 5, 2018). https://www.science.org/doi/10.1126/science.aam7240

7. Breitburg et al.

8. Annie Snider, "The War Over Chesapeake Bay," *Politico*, May 25, 2016. https://www.politico.com/agenda/story/05/obama-chesapeake-bay-restoration-000127

9. Alexander Kaufman, "Trump's Agriculture Pick Vows to Fight for Chesapeake Bay Cleanup Despite Budget Cuts," *HuffPost*, March 3, 2017.

Ephemera

KULVINDER KAUR DHEW

Ephemera, 2014. Charcoal on paper, 18 × 26 inches. Copyright © Kulvinder Kaur Dhew.
Private collection, New Jersey, USA.

Calling the Kings Back up a River They Lost

MAYA KHOSLA

Salmon can detect water-borne concentrations as small as
50 ppb. In martini equivalents, that's roughly one drop of
vermouth in 500,000 barrels of gin.
—TOM JAY

Then we carried crushed fish bones to the headwaters,
tossed them in with ash. All the cautious optimism,

all the longing from 64 missing years, tossed them in.
Winter wrens making their way through the carvings
of undercut, a great mass of root-wad. One of us worrying
the timing would be off, the dilution factor of bone-salts
in water would come in under fifty parts per billion.

Not saying it. One of us humming. The Willapa salmon might know
to migrate the miles, the tender measure of dilution, to swim
vertically up the falls. To run a curving line, determination
despite the turbulence of tripled flows, bound for beds
of gravel. Already rain and all the smaller inflows stirring up

a cloudy mud cover of safety, the gardens of fallen branches
going under. Riffles showing white seams, water breaking

open, voices over gravel. Already the dark tea of leafy debris.
All of it good, good enough to hide the passage of salmon.
Their surge catalyzed into action by the blend of smells.
Ash and bone, oil of roe, joining the leaf-rot and silt. The body
a vessel—only heart and kidneys remaining, shoved to corners

to make room. Loyalty, a new generation the fish would not see,
holding them as one river opposing the violent onrush of another.
The body burning itself down, down, to the essentials of racing.
Nothing but the drive, dorsal bones showing through, the thunder
of tails hitting gravel—a reflex action driving off the silt.
One fish sliding, blood-red, past a sudden band of sun.
The silvers of memory never weakening. Bent alders and willows,
the interlacing fingers of shade, dipping in. As mitigating forces.

As a way to join the silica, the nitrates and phosphates,
the language, the crazed and pounding language, of return.

Taking Measure

~~~

AKIKO BUSCH

BY 7:45 IN THE MORNING IT'S IN THE LOW EIGHTIES, AND A HAZE HAS
already settled over the Hudson, signaling the heat of the coming day.
My friend Michael and I have come down to the Newburgh pier in an
effort to gauge the direction and speed of the river's current. As tranquil
as it seems on this July morning, it is an unquiet river, I know. This area
of the Hudson is a tidal estuary, and the direction and velocity of its
tides and currents are ever shifting. So variable are they that the Native
American name for this river was Mahicantuck, "the river that flows
both ways."

The river's changeability is of interest to us now because the following
day over a hundred swimmers will gather at this spot for a cross-river
swim. As volunteers who help to arrange the swim, we need to know
when the ebb tide is slowing, when it slackens, and when the swimmers
can make the crossing safely, with as little downstream pull as possible.
Although the tide charts issued by the National Oceanic and Atmospheric
Administration help to make these predictions, there is something about
taking our own tidal measurements and being immersed in the current
itself that allows us to fine-tune NOAA's calculations.

Michael and I get to the river near the end of the ebb current cycle.
We've brought with us a measuring device called, improbably, a global
flow probe, an instrument that computes the speed of the water as it
travels downstream. It is a simple anodized aluminium tube marked
with depth measurements in both inches and centimeters. On one end
is a cylinder, inside of which is a small propeller that the river water
will surge through at a certain speed. Although it looks like part of a

child's toy, the propeller is, in fact, equipped with sensors calibrated to gauge the average water velocity in feet per second. On the other end of the tube is a small screen that displays the readings. Plunged into the water at intervals of five minutes, the monitor will tell us how the tide is ebbing, when it has arrived at slack, and when it turns to flood.

It's good information to have, but there is another reason I look forward to my appointment with the current. I value this exercise of comparing my own sensations of the river's tidal force with the readings on the monitor's display screen. Scientists often speak of the ground truth, what their own ears and eyes tell them as opposed to what appears on pixels, sensors, satellite imagery, infrared photos, or whatever other remote data information system happens to be in play. What we think of as knowledge so often now seems to require its own calibration, some manner of settling up remote data with those things we experience ourselves. Slipping into the current this morning seems a good exercise in that kind of reconciliation. Out here on the river on a July morning, it is the water truth I am after.

It's been a wet summer and heavy rainfall upstream can influence tidal flow. I anticipate an ebb current with a strong and steady pull, but of course that is not what I feel when I slide from the pier into the gray river. The water temperature is in the low eighties. I let myself float, but only moments later, looking back at the pier, I realize I *have* been carried twenty feet down river. At 8:11, Michael calls out from the pier that the water velocity reading is 0.42 feet per second. I just say it's a strong draw, and that also goes into the notes he's taking.

I know, too, that wind has a strong influence on the behavior of surface water and that a south wind can subdue the effect of the ebb tide. This morning, though, it is only a light breeze from the south that blows up the river, scalloping its surface. If you were watching the water from the shore, you might think the tide was coming in, but even as the ripples wash over me, I can still feel the downstream pull. Rivers tell you one thing, and then another. By 8:20, Michael tells me it measures at 0.36. A little easier, I call back to him, treading water.

I have learned that when the tide switches from ebb to flood, it tends to progress outward; that is, it begins at the shoreline and ends in the middle of the channel. But the pier I am swimming from extends 100 feet from the shore, a calculation that adds another variable when the direction and velocity of the water shift. The surge of water can also vary from one to two feet per second in the period of a single minute. And I know as well that flow and velocity are different things. The first has to do with the volume of water, the second with its rate of travel, the number of feet it moves per second. All of this is why I can't help but think what folly it is to try to be precise in estimating the river's current. I may as well try to be accurate about the color of the water, the angle at which the light falls across it, or the distant taste of seawater that lingers in the estuary.

For the next half hour or so, Michael calls out the readings to me: 0.34, 0.28, 0.21. And I respond by telling him how the current feels, using a different vocabulary: strong, easy, easier. "Nothing to it," I tell him when I feel the tide slowing, and a few minutes later, "a piece of cake." His concise register of small numbers and my own improvised impressions make for a comedic dialogue, going from shore to water and back. Still, it seems an increasingly stilted and abbreviated exchange, inadequate to capturing the sweep and strength of the wide river.

What we are after is nothing but predictions, imperfect conjectures about the future. And I know it is an inexact science, this business of trying to understand how quickly things around us are moving, changing. I take in the feel of the water, what is visible on its surface, and the numbers that the little whirling propeller sends to the monitor's display screen. The information shifts in and out of alignment. None of it perfect or final. It arrives in bits and pieces. It's 9:10 now, and the average velocity of the water measures in at 0.06. Nearly slack before flood. The river isn't taking me anywhere, but you'd never dream of saying it is still. Again, I find myself amazed by all the different strategies we have devised in trying to gauge those forces that so influence our lives. I am adrift now, floating on my back, the light and the water spilling over me in equal parts.

Giant Tuna

Maya Khosla

Bending low to carve an ice-fish, the fisherman hears
the beating of a distant drum. His heart still

at sea. Left thumb long gone to a notion of danger.
The sensation of a digit where there is none.

He could feel the lightning, the fishing line burning
through his hands, seconds after he noticed his hook

tugged under waves. Then the lurch, the air flying past,
the cold snap and pulse of blue upon blue gathering him

into its salty arms. Fists of water rising over their infinities
punching his lungs. The mind's currents whistling

without thought. The line sinking too fast
for the instinct of releasing hold to translate

into releasing hold. The blue shadow below
deepening, the yellowfin tuna continuing

to breathe through movement the way they always do.
Mouth wide, gill rakers wide, tail undulating.

To stop means to perish. Crimsoned muscles
drive themselves forward, awake, asleep, awake.

The fish, the line, dragging him far. The men on deck
waving, shouting directions. Nothing he could hear.

Notes on an Impure Hydropoetics & the Water Molecule

BRENDA HILLMAN

IT IS GOOD TO BE THINKING ABOUT WATER MOLECULES. IT WAS ONCE thought that earth's water came from asteroids, comets, or protoplanets. But studies of moon material and of the asteroid Vesta—one has to love the name—have shown that the earth was probably formed with a lot of water already present in the minerals. The dwarf planet Ceres, alone in her icy path, has a great deal of underground water. Mostly Earth's water was squeezed out of rocks as if by some giant in a European fairy tale. There was a great explosion that created vapor from its opposite, volcanoes, with atmosphere condensing to form rain and to make the oceans. The oceans boiled, then they cooled, then single cells that didn't breathe, then single cells that did breathe, then eukaryotes, then flowers, then Joni Mitchell writing songs about grilling salmon by the ocean with the tide rolling in with her dysfunctional love in Malibu. i've been inspired in my poetry and political activism by thinking about water during a drought, to engage in a spiritual and sympathetic way that will show respect and increase relationships with the world. There are entities active on multiple planes of reality, either in philosophical ways or in what Van Morrison calls *astral weeks*.

~

West Marin stream.

> . . . she hears a small splashing,
> memory of *when*, cherished
> sound of *him* in an inflatable
> pool, naked, singing,
> *I'm naked*, he sings, *Naked, I'm naked tonight*
> —C.D. WRIGHT, "WATER BABY POEM"

≋

BRENDA HILLMAN

i asked loved ones about their first memories of water. i asked my San Francisco–born husband this question and he said: fog. i asked my parents for theirs; my father said, being baptized in the Leaf River (near McLain, Mississippi). My mother said, diving off a rock into the Guaíba River in Porto Alegre. My father came from southeastern Mississippi Scotch-Irish subsistence farmers, hard-headed and smart and not always charitable. My mother's people were immigrants to Brazil in the nineteenth century, rebellious Protestants determined to save Catholics from idolatry and to teach reading so the poor could read the Bible without intervention by priests. They stayed there as teachers for multiple generations, living at the edge of poverty. All bodies are immigrants, all nations and bodies are melting pots and melting plots. A confluence of destinies flows inside every human body. Every body is a watering place. How did you feel when you jumped? i asked her. "I said to myself, 'Jump, Helen!'" she remembered ninety years after it happened.

~

A poetics of impurity in a time of degradation of species, when the water of so much of the world is undrinkable. Tap water runs brown in some parts of Los Angeles. People must use bottled water and add plastic to the oceans. Dead zone in the Pacific the size of Texas. People can set fire to the tap water in some communities of California because of fracking. Unthinkable to have undrinkable water in a country that can send a car into outer space. Failure of empathy, failure of an economic system. Should the work of poetry be only ceaseless mourning? It should be intolerable to every living person that women in Burundi have to carry water from so far away they are at great risk to themselves walking through the dark. Their other choice is to use bottled water—if they could even afford it. Plastic pollution from poverty. Snapper choking on the rings of BudLight. To live in ceaseless mourning but not only.

〜

Fady Joudah in "Revolution 2" writes:

Imagine we desisted to vote
where and when we could

how would the poem
proceed? You'd say

We'd be asking for water to die
But you Sir are no water

〜

In the mid-fifties, i flew with my family over the Amazon to Brazil where we would be living for a few years. In those days women and little girls wore white gloves on airplanes and the stewardesses wore white gloves serving the food on trays. i want to say the plane may not even have been a jet. Our father was going to advise farmers and my mother was returning to the country of her birth. My mother told us to look down from the airplane: the dark waters of the smaller Rio Preto were flowing into the larger Amazon basin. i was tiny but the memory is clear. That sense of confluence—of the less likely flowing into the more likely—would correspond to my exploratory poetics—word-inanimate-mystery-justice—things of this world in a ceaseless flowing and i've always been suspicious of stasis, camps, and rigid ideologies.

〜

BRENDA HILLMAN

The symbolism in the dream—water is the unconscious, water is birth. That water is as the "bloodstream of the biosphere" (Johan Rockström's phrase) is a fact we forget daily as we use three gallons of the precious stuff for brushing teeth. The sun may be the main instigator but the water is responsible for the flow of energy. From their water-filled worlds, my parents settled in the deserts of Arizona. i grew up preferring dry hot things to cold and moist. Desert children are taught the value of precious water, even in Tucson where, in the '50s, the golf courses were increasing.

Tucson saguaros.

NOTES ON AN IMPURE HYDROPOETICS & THE WATER MOLECULE

Those thorny machines of saguaro get by on so little water. In several communities at the outskirts of the city Tohono O'odham, (Pima) peoples whose homes, made of adobe, with thorny ramadas and fences, practiced traditional agriculture using annual monsoon rainfall and conserving carefully so as not to deplete the aquifer. In California we are repeatedly coming up against drought and wildfire—calling for a wisdom that is not optional.

∾

Writing about the earth's elements has been a voyage through an "impure" style from romanticism to modernism and postmodern ritual. In the '80s. after a decade of studying alchemy and Gnosticism, i decided to write about the elements—earth air water fire—from a perspective of an experimental California poetics. At the time, innovative poetry of the West Coast included Language poetry, the remnants of the Beat traditions, and various forms of lyric experiment; few were taking an ecological approach. Ecopoetics was a new term. i had often been inspired by the San Francisco Renaissance poets and the powerful West Coast nature writing tradition, but they mostly came from a male perspective. After *Cascadia*, a book originating in experiments with California geology, i worked on air, water, and fire for a total of two decades. Information, research, science, and the internet made the poetry mottled and impure. The weather grew increasingly intense with climate change; "normal sentences" broke in fragments, pushed against the right margin. When it was water's turn in the aughts, i visited various hydrologic regions and rivers to think about the water molecule as a concept and a unit. My method with each element was not to write "about" earth or air or water, but to invite the element into the poem. The more i learned of water, the more excited i became about hydropoetics, including information and research gathered both from scientific sources and also by intuition and trance.

California live oak.

≋

Kim Hyesoon, in "The Story in Which I Appear as All the Characters" (trans. Don Mee Choi), writes:

> Even though the memory vanishes
> like water that has been gulped down
> the song always remains!

≋

The physical properties and metaphoric implications of the water molecule enchanted me. i developed a crush on the water molecule, its poetic tumbling freedom and wildness, third most popular molecule in the universe, in charge of the atmosphere, precipitation, respiration, and the life cycle. The pronged 2 H's and the O; the slightly negative and the slightly positive attracted to each other; the amphoteric (both acid and base) properties; colorless but a bit blue; the best solvent. It has been called oxidane, hydrogen oxide. i began to think it has a special kind of thought. It isn't that nonhuman objects don't have experiences, it is just that we aren't supposed to overwhelm them with ours. In my childhood animism, i knew the water inside a cactus might be experiencing something different from a brenda. The fate of the molecule was to tumble between watersheds with athletic poetic force. For the hydrologic regions, i used slashes to indicate dams the molecules have to leap over. To develop an anarchic approach to document and research, to begin from fragments of consciousness and feeling rather than a pure idea. Will ecological poetry help the world's water crisis? No. But poetry has concentrated amounts of empathy and imagination.

≈

If in falling   rain names what it touches

If beneath the tree   a dry radius describes

form   steps forward   wearing its suit   of summer's dust

    —BRIAN TEARE, "LARGO"

≈

BRENDA HILLMAN

The self is also watery. When writing about ecological matters, to hold the idea of relationship that does not relegate human emotion to sentimentality; to enact but also to question the theories of a "whirling decenteredness". . . Reading Derrida and Ashbery in the '80s, but as a threatened woman, i thought, selves may be questionable, but they definitely exist. Reading and traveling California during the drought while the United States was invading and re-invading the Middle East took the self out of the self and allowed a different subjectivity; it further radicalized my writing and my other actions.

〰

uma certa paisagem sem pedras ou sobressaltos
meu salto alto
em equilíbrio
o copo d'água
a espera do café

(a certain landscape
without stones or shocks
my stiletto heel
balanced
the glass of water
the waiting for coffee)
 —ANA CRISTINA CESAR, "[WHEN BETWEEN US
 THERE WAS ONLY]"

〰

In 1900, nearly one third of the hundred million acres of California were under cultivation and it was largely monoculture farming, and the lack of diversity in much of this agriculture has been terrible for the water table as well as for the soil. Grazing required the fast-growing Europeanized grasses. Specifics about land use can make a poet very depressed, thus one can maintain focus on water sources: the mighty rivers far up north, the underground streams beneath the lava beds, the ice caps of Shasta, the thin vast tumble of waterfalls in Yosemite to the reservoirs that hold our tap water. Some were saying every almond requires five gallons of water, but farmers deny it. Still, according to the Pacific Institute, "agricultural water use could be reduced by . . . about 17 to 22 percent, while maintaining productivity and total irrigated acreage." i dream of the days when farm workers will share in the profits of big agriculture. http://pacinst.org/making-conservation-california-way-life/

≈

The sky song is a blues the sea
comes into on repeated lines.
—CAMILLE DUNGY, "ARS POETICA: COVE SONG"

≈

For a while i was going to DC to protest wars by visiting Congress. i sat in congressional hearing rooms where funding for wars in the Middle East was debated. Because i was working on *Practical Water*, i devised brief occult rituals with water as part of my activisms. The concept of the water molecule took on new life in these official settings. Sometimes i would pour a small amount of California water on the floor of the hearing room, sprinkle water on guns in the military displays, or work in trance while squinting, concentrating on the digestive tracks of senators making the war policies. As has been noted, the polluted Anacostia River near the Capitol has to cope with congressional shit. My practices during direct action are metaphoric and nonviolent. Impure activism: single actions have little effect but that doesn't mean they are useless. Collective actions can seem useless but mighty. The revolution is far away but get your ass outdoors when there is a threat. Give up perfectionism. Do more as you are able but don't make yourself sick. Get hugely educated. Take your imagination with you. If it doesn't feel uncomfortable, it doesn't count.

≈

Bob and i went to Libya with poets Forrest Gander and C.D. Wright, where we visited the covered village of Ghadames at the edge of the Sahara. It resembles the commune of southwestern pueblos but it is a covered fortified structure that is a maze of inner streets and small family dwellings. Several hundred families lived there at any one time, and there were five water sources arising miraculously from the Saharan artesian wells. We walked through astonishing hallways underground, shafts of light coming through. In the subterranean canals, water was managed communally and irrigation was regulated according to a complex time system that involved tying knots in palm leaves. It was hugely inspiring. In that system, the water keeper made sure the jug was filled each time from the underground stream. A hole in the container released overflow water into the canal. The time it took for the process—about three minutes—represented a unit of measurement, one kadus. Each kadus was noted by a knot in a palm leaf, and regulators told how much each house or district got. After each household got their amount, a stone was placed over the outlet. People could find out both the time and the amount of water that had passed by visiting the log in the main square. A time and water management system so admirable in this time of greed and destruction. Sometimes when i am very sick about what happened to Libya, and i suspect it will be a very long time before an international meeting of poets occurs again, i think of that water system like poetry itself, and the time and water keeper as the poet with some idea how to get exactly the right amount.

Seaside Hallucination

WILL MCGRATH

DURBAN, SOUTH AFRICA

I shucked my clothes on the beach and ran toward waves. That day was windy gray, but close to shore the waves were warm and wild, the Indian Ocean detonating around us: bodies brown and beige, my body pink and rare like an underdone chop. We were in our underwear. Men in soaked briefs, women in brassieres and yellow shower caps, or with bright red plastic bags wrapped over their hair, beaming there like stoplights. Further out the Blacktips and Tiger Sharks slid in ceaseless rolling motion, and all of Durban bobbed—beautiful bellies pointed toward the sky.

The ocean snagged my legs and I somersaulted, tumbled into three Sikh men I'd seen earlier on the beach—pensive, or so they'd seemed to me, their faces carved with worry. The four of us came up laughing, rubbing salted eyes, and then we all were ocean-bowled again. We breached and sang out that same shocked laughter, made giddy as we were unmade by the sea. Their dark beards were plastered to their chests, turbans soaked and coming now undone.

The ocean breathed around us because the tides are always breathing and the seas are lungs. Wasn't it Kepler who claimed Earth sweated, farted, secreted earwax?—so of course it breathes: that slo-mo planetary inhale-exhale. The seas are global lungs and Earth's nerves are made of mushrooms, a fungal network knit beneath, like that subterranean shroom in Oregon, three miles across, Earth's largest living thing—but that's not the point—the seas are lungs, this isn't a metaphor. Most of the oxygen we've ever breathed and ever will breathe has been huffed into the sky by sea-dwelling phytoplankton, tiny algae, diatoms and dinoflagellates, those microthings that honed their breath across the eons. This was long before terrestrial plant life was a twinkle in the cosmic eye.

So yes, the seas are lungs and they practice ujjayi breathing, hatha yoga's measured mindful two-step, slowly scraping in,

then out,

against the back of nose and throat. *Oceanic breathing* people sometimes name it, a sheathing sound, the sound of sea within you.

In Durban we were tumbled in the waves, all our tender bodies knocked about in salty joyful fulmination. The ocean was gun barrel–colored, sharkskin-colored, and in that moment it seemed we might all be accepted by the sea, we could feel the webbing deepening between our fingers, feel it stretching toe-to-toe, we could dwell forever in this churning, we could be the churn.

SKELETON COAST, NAMIBIA

Sam was five and Eve was three and we were touring down the Skeleton Coast, Ellen at the wheel of a rented jeep, the four of us in thrall to the otherworldly artwork of the seaside. Here the ocean is studded with half-sunk tankers and the ruins of ancient rusting barges, thousands of wrecks beached along that desolating shoreline. They were ensnared across the decades in devilish currents and walls of fog—the vaporous offspring where frigid Atlantic and warm Namib winds conjoin. The hull of the *Eduard Bohlen* is sunk there in sand like the carapace of an enormous overturned pill bug. It ran aground in 1909 and rests more than one thousand feet inland now, for the border between Atlantic Ocean and Namib Desert is a shifting and imaginary thing.

As we drove, I thought of Christo and Jeanne-Claude—that husband-and-wife pair of art world sprites—who in 1969 played a trick on the Australian seaside and made it disappear. Near Sydney, they veiled a mile and a half of coastline in fabric: up and down a team of tiny human yo-yos went, professional mountaineers who rappelled the cliffside and packaged the rock face in one million square feet of woven material, lashed down the shroud with thirty-five miles of rope. The result was a surreal melt of land into sea. When the wind ran through the battened fabric, the cliffside became at once both liquid and solid, thrumming with unfathomable purpose, an echo of the ocean.

Their greatest trick was the simplest: by concealing, Christo and Jeanne-Claude revealed what was there. For ten weeks the coastline remained in packaged form, then was unwrapped. The art was gone, or maybe changed its state. They'd wanted to sculpt something no one could own, said their work dealt in ephemerality, in temporariness—creations that took years, decades, of planning and execution, only to be dismantled weeks later. "Is art forever?" Christo asked. "It is a kind of naïvete and arrogance to think that this thing stays forever, for eternity. All these projects have the strong dimension of missing, or self-effacement, that they will go away, like our childhood, like our life."

Sam and Eve existed for a golden flashing moment of their childhood among the dissolving sculptures of the Skeleton Coast. Having fled the jeep, they now cartwheeled in the sand and jitterbugged before the ruined fishing trawler *Zeila*, which in 2008 ran aground in Namibian coastal fog. That boat's bones are now a bird perch, apocalyptic backdrop for my children's sandy stunting. These sculptures were placed by artists whose names we've forgotten, artworks mired for now between the earth and sea. The Atlantic will reclaim them though, the sea unwrapping packages, patiently returning us to our component parts.

MICHIGAN, UNITED STATES

I'm time traveling now as I look through these pictures, reaching back to former seasides, attempting to grab hold of things that will never again be, since we are older now, and the kids are older now, and now older than when I began this sentence, began this thought. So I'm time traveling to a shore where Sam was still a toddler, not yet two, and Eve and Mara hadn't yet become.

I held him in my arms and from behind a window's safety we looked across that glacier-carved and raging inland sea. A storm was moving over the water. It was like that scene from Hokusai, the woodblock waves in Prussian blue, the cresting whitecaps with their greedy clutching fingers. I felt Sam's body tighten as we noticed a boat making for safe harbor.

"It's dangerous," I said, perhaps to him.

He asked me, "Why?"

"Because the storm could knock that boat right down beneath the waves."

We tracked the curious progress of the boat that seemed a bathtub toy, took in the grandeur of the gale.

After a while he spoke.

"There people?" he said quietly. "That boat got people?"

CALIFORNIA, UNITED STATES

The past is an ocean, says Maggie Smith, it is
> a tide that drags out
> but won't return to shore: even your question
> has been carried off. Look, you can see it floating.

Not far from where I swam in Durban, five men were carried off into the Indian Ocean during a baptismal ceremony, died as they were trying to be born. Near Cape Town, the Atlantic took three men from a similar rite, raptured by the sea. The Pacific Ocean claimed a man named Benito Flores from a sacramental immersion in Santa Barbara County, California. This baptism had been taking place along the same stretch of beach where DeMille filmed *The Ten Commandments*, a cinematic stand-in for the Red Sea, which has precedent—it has been said—as a tool of murder by the Abrahamic God. Flores was sucked out in a wall of water and his body washed to shore two weeks later.

Up the coast from DeMille's biblical beach, we visited a boardwalk amusement park one night. Eve was an inch or so too short to ride the swings but was desperate for it, so she and I bent ourselves to conspiracy: foreheads pressed, I whispered incantations, took surreptitious glances at the teen attendant. *You're big enough*, I told her, and she repeated it. *Don't look at the man, just walk*. With a breath, Eve centered herself, reined in the wild cantering of her body, and we blew past—her head drawn tall, her face a pantomime of adult gravity. That somber ticket-taker never gave a second look, brow furrowed to his task of adolescence, newly sprouting mustache as we passed.

Saucer-eyed and cleaved to me, Eve was bathed in neon. We chained ourselves into the swing's metallic harness, strung from a glowing mushroom top, an electric toadstool sprouted from subterranean loam. Then into that beachy night we were lifted. From our higher vantage we tried to parse the hieroglyphics scrawled in lights along the pier, but all was churn and flow and tides of people. Out beyond the boardwalk hulking freighters skulked the line. I could see how they would all be rendered sculptural, a gentle demolition set against the rhythm of that steady planetary breath. Could see how this would all go on without us. Sam was waving up at Eve and me, too scared to ride the swings but delighted by his little sister's joy. And there was pregnant

Ellen, full of Mara, Mara who was not yet Mara, or maybe was, herself suspended in some shifting inner sea. How mathematically improbable that any of us should be—how unlikely and miraculous. Then the gears engaged and

we were spun over dark water
            legs aflutter
                        draped across unfathomable
                                    alien deep

Acequias as Quipus, Quipus as Poems

ARTHUR SZE

FOR THE LAST FORTY-FIVE YEARS, I HAVE LIVED IN NORTHERN NEW Mexico; during this time, I've been involved with two acequias: the Ancon de Jacona and the Acequia del Llano. The word "acequia" is derived from the Arabic *as-sāqiya* and refers to the irrigation ditch, as well as the association of members connected to it, that transports water from a river to farms and fields. For eighteen years, I was involved with the Ancon de Jacona, twenty miles north of Santa Fe, and, for the last four years, I have been actively involved with the Acequia del Llano.

The Acequia del Llano is the youngest of the four acequias that run through the city of Santa Fe. It is one and a half miles long and begins at Nichols Reservoir and runs eventually into the Santa Fe River. Fourteen members are in this ditch association, and the acequia irrigates about thirty acres of gardens and orchards. Some of the endangered and threatened species that draw on this watershed include the southwestern willow flycatcher, the least tern, the violet-crowned hummingbird, the American marten, and the white-tailed ptarmigan. Other wildlife in the area include deer, black bear, coyote, turkey, and quail.

As a *parciante*, or voting member of the acequia, I am actively involved in the maintenance of this ditch. Just last week, on April 7, all of the members came, or hired workers who came, to do the annual spring ditch cleaning; this involves walking the length of the ditch, from reservoir to river, cleaning, with shovels and clippers, branches, silt, and other debris that has accumulated over the late fall and winter.

The ditch association is organized with a *mayordomo*, ditch manager, who oversees the distribution of water according to each *parciante*'s amount of water rights and also verifies that each *parciante* follows the strict watering schedule. The current *mayordomo*, Mike Cruz, drives periodically up and down Canyon Road and monitors the usage to make sure no water is being wasted and spilling into the street. The rules of the acequia are that if someone upstream is not using water during their allotted days and times, someone downstream can draw on that water. However, one cannot draw on water when someone downstream has the right to that water. The *mayordomo* resolves any conflicts about water usage, and he also has the authority to fine land owners who do not abide by the rules. For our land, we have two days a week when we can draw water: from Thursday at 6 P.M. to Friday morning at 7:30 A.M. and from Sunday at 6 A.M. to Monday at 6 A.M.

The acequia runs at a higher elevation than all of the land held by the *parciantes*, so the flow of water is gravity-fed. On Thursdays and Sundays, I walk about a quarter of a mile uphill to the ditch and drop a metal gate into it. Water then backs up, and as the level rises, water goes down two pipes. Two large green holding tanks fill, and then water runs into a complex system of pipes set with timers, so sprinklers water grass, a garden, a few hundred-year-old apple trees, and flower beds around the house, according to the different zones and times. I manually turn on and move a secondary set of rotating sprinklers to water the orchard that has apple, peach, and pear trees.

New Mexico is a very dry landscape, and, this year, due to climate change, the drought is particularly damaging. By July 1, the annual rainfall is normally 4.5 inches, and, as I write today on July 5, the annual rainfall is an astoundingly low 1.25 inches. Severe water conservation measures are in effect throughout the city, and everyone who has the right to draw water off the Acequia del Llano is aware how precious this resource is, and the sharing of this irrigation water actively promotes community needs over a single individual need. It also serves as an important reminder and contrast to our consumer culture where, as Ezra Pound once observed, "[Nothing] is made to endure nor to live with / but [. . .] to sell and sell quickly." During the Depression, there are stories of impoverished families sharing sheep bones to flavor broth. Just like the water that was passed along the irrigation line, bones were passed from family to family to flavor subsistence soup.

Each year the irrigation season runs from about April 15 to October 15. In April, as I get up in the dark in the early morning and walk uphill to divert water from the ditch, I notice Venus as well as Orion and other constellations of stars. I see lights from a dozen houses on the far side of the Santa Fe River flickering in the darkness. Toward summer, I notice the constellations shift in the sky, and by July 1, when I walk uphill, I am walking in early daylight. By mid-September, I will again be using a flashlight to head uphill in the dark and will be listening for deer or coyote in between the piñons and junipers. Connected to this seasonal rhythm, I am in biweekly contact with flowing water and recognize, as a steward, the privilege it is to use it to irrigate the land. This liminal awareness sparked my poem, "First Snow":

FIRST SNOW

A rabbit has stopped on the gravel driveway:

imbibing the silence,
you stare at spruce needles:

there's no sound of a leaf blower,
no sign of a black bear;

a few weeks ago, a buck scraped his rack
　　　against an aspen trunk;
　　　a carpenter scribed a plank along a curved stone wall.

　　　You only spot the rabbit's ears and tail:

when it moves, you locate it against speckled gravel,
but when it stops, it blends in again;

　　　the world of being is like this gravel:

　　　　　you think you own a car, a house,
　　　　　this blue-zigzagged shirt, but you just borrow
　　　　　　these things.

Yesterday, you constructed an aqueduct of dreams
　　　　　and stood at Gibraltar,

　　　　　but you possess nothing.

Snow melts into a pool of clear water;
　　　and, in this stillness,

　　　　　starlight behind daylight wherever you gaze.

≋

If you take an aerial view and visualize the acequia running along the
hillside as a primary flow and then locate the many subsidiary or secondary flows running downhill, perpendicularly, from the main flow,
to irrigate various orchards and fields, you will see a quipu composed of
water. In Inca culture, a quipu is made out of spun fiber and is defined

ARTHUR SZE

~ 52 ~

in *Merriam Webster's Collegiate Dictionary* as: "a device made of a main cord with smaller varicolored cords attached and knotted and used by the ancient Peruvians (as for calculating)."

The word "quipu" has two spellings: the older version, "quipu," is based on the Spanish spelling for the Quechua word that means "knot." It is more often spelled "khipu" today. In either case, the quipu is usually made of cotton; it is lightweight, portable, and encodes information. There are two kinds of quipus: numerical and nonnumerical. The numerical ones record accounting information. Researchers have examined the knotting in quipus, and they are able to read the numbers, one to nine. The location of knots on the numerically-based cords follows a base-ten decimal system, so a quipu can easily incorporate ones, tens, hundreds, and thousands. For instance, a quipu that accounted for containers of potatoes in a mountainside storage vault would be invaluable during a famine. The rulers could pull potatoes out from storage, feed people, and retie the knots, so that inventories of food could be kept up to date. Numerical quipus track quantities of sandals, gold, census numbers of people living in a village, and so on, so these quipus provide essential day-to-day information that helped the Inca rulers respond to crises and rule the kingdom.

Interestingly, poet and translator Brenda Hillman has mentioned that water usage in the Sahara was tracked by tying and untying knots. And there is also evidence of ancient Chinese quipus, composed of silk, that predate Chinese characters. In the next-to-last chapter of the Dao De Jing, there is a passage that says let people go back to communicating through knotted cords. In a poem, "Thoughts," by Du Fu, in the Tang dynasty, the speaker sits on a porch at night and muses how people struggle futilely for glory, and then the speaker considers how someone started communicating by knotting cords and now there's the mire and endless bureaucracy of government! So the use of knotted cords as a vehicle for communication can be found in many ancient cultures.

Focusing on the nonnumerical quipus, the possibility they encode language is tantalizing, but, so far, no one has been able to "read" or decipher them; the knotting does not follow a linear base-ten accounting system. Nevertheless, several important historical accounts provide evidence that quipus incorporate narrative information that encodes the myths, legends, and histories of the Incas. One important historical account describes an Inca runner who arrives at a remote mountain village. He pulls out a quipu, and a *quipucamayoc*, a specialized reader of quipus, looks at it. The village then begins a revolt against the Spanish. Here, then, is evidence that the quipu must have contained narrative information.

As a poet, the nonnumerical quipus interest me most. I like to think of the knotting in a quipu as a physical and visual reminder of the way a word or phrase may be repeated and turned in a poem. Robert Fitzgerald once remarked that repetition in Homer utilized elegant variation, that each time the repeat occurred, the word was enriched and deepened in meaning.

In composing my book, *Quipu*, I looked at all the dictionary definitions of the word "as" and noted ten different meanings. In the title poem "Quipu," a sequence of nine poems, I took it upon myself to keep using the word "as" and, over time, allowing different meanings to accrue. This was an organic way to layer, deepen, and simultaneously enrich the meaning of the poem. The poet Cole Swensen has noted that repetition frequently involved casting a spell. I agree and would add that repetition can also become a form of insistence. The knots can refer to repeats of words, and they can also refer to repeats in syntactic patterns; moreover, as in a quipu where the spin of fiber in secondary cords can shift from clockwise to counter-clockwise, I envisioned a shift where nouns could turn into verbs. The sequence "Quipu" is too long to reprint here (the complete poem was first published in *Conjunctions* 35, "American Poetry: States of the Art"), but some of these thematic currents are clearly initiated in the opening section:

Quipu

I try to see a bald eagle nest in a Douglas fir but
catch my sleeve on thorns, notice blackberries,

hear large wings splashing water in a lagoon.
I glimpse a heron perched on a post above a tidal flat,

remember red elderberries arcing along a path
where you catch and release a newt among ferns.

And as a doe slips across the road behind us,
we zigzag when we encounter a point of resistance,

zigzag as if we describe the edge of an immense leaf
as if we plumb a jagged coastline where tides

wash and renew the mind. I stare at abalone eyes,
am startled at how soft a sunflower star is to touch,

how sticky a tentacle of an anemone is to finger.
When we walk barefoot in sand, I sway

to the motion of waves, mark bits of crabs
washed to shore, see—in an instant a dog wrenches

a leash around the hand of a woman, shatters bones—
ensuing loss salamanders the body, lagoons the mind.

The flow of water is also the flow of language: water and poetry are essential movements that affirm and shape life. Water can be conceived of as beginningless beginning and endless end; if water has no shape of its own, it can take any shape and has infinite possibility. In poetry, I am interested in a finite thing that has a multiple or polysemous range of expression and meaning. Poetry utilizes a finite set of words and yet has the possibility of reaching into the infinite, and it calls our attention to the mystery of existence. I am reminded of Dogen's dictum, "Water is the koan of water." If water is a riddle of itself, then all of creation is mysterious and marvelous. It is astonishing life exists and that there is anything at all.

In looking back over my time in New Mexico, I realize that the scarcity of water has helped me pay close attention to living details and the profoundly changing landscape. This realization is embodied in this poem set in Jacona:

RED BREATH

Shaggy red clouds in the west—

unlatching a gate, I step into a field:
          no coyote slants across with a chicken in its mouth,

          no wild asparagus rises near the ditch.

In the night sky, Babylonian astronomers
          recorded a supernova
          and witnessed the past catch up to the present,

          but they did not write
                    what they felt at what they saw—

they could not see to this moment.
From August, we could not see to this moment

    but draw water out of a deep well—
    it has the taste of

    creek water in a tin cup,
        and my teeth ache against the cold.

Juniper smoke rises and twists through the flue—

    my eyes widen
    as I brush your hair, brush your hair—

    I have red breath:
    in the deep night, we are again lit,

        and I true this time to consequence.

Another time, walking in the ditch of the Ancon de Jacona before spring cleaning, I marveled at the landscape around me and, in imagination, moved across space and time. The village of Jacona is in an agricultural valley fed by the Pojoaque River, and Los Alamos, the birthplace of our atomic age, sits visibly on a mesa to the west. The following poem is in an invented form. Starting with the title, each line picks up a word or words from the previous line, and, at the end, a word is picked up, again, in the title, so my form embodies line and circle. In a similar way, an irrigation ditch in northern New Mexico that carries water to the fields forms line and circle; it marks the seasonal beginning and end that supports and sustains life here.

Sight Lines

I'm walking in sight of the Río Nambe—

salt cedar rises through silt in an irrigation ditch—

the snowpack in the Sangre de Cristos has already dwindled
    before spring—

at least no fires erupt in the conifers above Los Alamos—

the plutonium waste has been hauled to an underground site—

a man who built plutonium-triggers breeds horses now—

no one could anticipate this distance from Monticello—

Jefferson despised newspapers, but no one thing takes us out of
    ourselves—

during the Cultural Revolution, a boy saw his mother shot by
    a firing squad—

a woman detonates when a spam text triggers bombs strapped to
    her body—

when I come to an upright circular steel lid, I step out of the
    ditch—

I step out of the ditch but step deeper into myself—

I arrive at a space that no longer needs autumn or spring—

ARTHUR SZE

~ 58 ~

I find ginseng where there is no ginseng my talisman of desire—

though you are visiting Paris, you are here at my fingertips—

though I step back into the ditch, no whitening cloud dispels this
    world's mystery—

the ditch ran before the year of the Louisiana Purchase—

I'm walking on silt, glimpsing horses in the field—

fielding the shapes of our bodies in white sand—

though parallel lines touch in the infinite, the infinite is here—

Atlas with Shifting Edges

Elizabeth Rush

So it was in summer again the loved ones went out to the
sea at a quarter to dusk

The part of them that could do nothing did nothing
& the light of them walked along

Walked west forgetting not the horror but forgiving . . .
—Brenda Hillman, "Poem for a National
Seashore"

THIS ESSAY IS AN ACCOUNT OF THE DAYS I SPENT DRIVING THROUGH
the Pacific Northwest while on a tour for my book *Rising*—a time of
wildfires, loss, and possible futures.

MILE 23.6

This morning in Oakland, I am thinking of Arthur Sze in New Mexico
and his water ditch. I can almost see him walking toward the water
ditch's gate. He lifts the latch to let the water flow into his cisterns so he
and his wife might bathe, might water their plants, might make tea and
wash their dishes. Water is both noun and verb. Something solid and
some human motion.

I won't try to paraphrase the poem he wrote. Know it has none of that,
no bathing or making tea. Instead there are magpies, and straight
edges, and circular saws. Dandelion stalks, peanut butter, rat shit, and

PVC. Basically, everything that could be in the poem is in the poem. It's there because of a deliberate act. Arthur woke, walked toward the water, and lifted the gate.

MILE ⟿ 77.7

In this exact cove just south of Limantour Beach a decomposing blue whale, the largest mammal on earth. I encounter the baleen first. Jet black and so shiny it seems it could be plastic. When I see it I think: insides of an air conditioner. But then I draw closer, reach out to touch what is there. The pliant pieces each ending in a flutter of hairy strips that bend but will not be separated from the rest.

The whale's labor does not end when it dies. The matter must go into the guts of the coyotes who come down here to feast. Must dissolve in the saltwater and wash out. For now there is a little something left, some vague body, syrupy yellow. Soft slag heap. I suppose this is the part that smells. I have not been brave enough to scan the mass in search of the animal's eye. But I did touch a giant vertebra picked bare and said aloud, "You were a beautiful whale."

A man in a straw hat and a woman in a vest and two kids—theirs, I suppose—tell me that this young whale was struck by a shipping boat, its spine severed. The sky is suddenly shot through with birds and the people all look up in unison. *Love, we will raise our children in a world where this happens.*

The tide rises and the waves around the whale carcass make momentary puddles of foam that turn blue, turn black, then sink into the sand. Littoral, one of my favorite words in the English language, in part because it sounds like literal. Literal means being interested in the thing itself, perhaps even in physical matter.

Littoral is an in-between place where land meets sea, where language meets orality. Currently my back rests against a bleached piece of driftwood, the trunk running perpendicular to a vertical cliff the color of dried wheat. It hems in the beach, the whale, and me. If I sat here long enough I could watch all this literally disappear beneath an ocean made heavy with water from the poles.

In that ocean, seals. The sea is so clear that when they dive down I can watch their bodies go all S-shaped. There is a kelp forest offshore and sometimes I mistake the bulbous ends of dulse for a seal's slick face.

People stop to take photos of the rotting body because we don't know what we are supposed to do. Before I walk closer I put on my sandals. I rise but cannot find the eye.

Woke to: mother and child deer grazing above my camp on the nearby hillock, its slope separates the whale body and me. A dune-colored hummingbird, heading south. Bunnies rustling in the underbrush. The click of bird claws on the picnic table, bees checking out my honey. Woke up already arrived. Dew on the rain fly and the ocean hissing beyond the cliff.

Wild peas—pink and
flame entwined with
the pines they rise
upon.

I read that in Puerto Rico, after Hurricane Maria, Dalma Cartagena taught her students to grow vegetable plants from seed. When they touch and tend to the sprouts they begin to recall that despite the storm's violent rending, they are always also part of something that is protecting them, nourishing them.

ELIZABETH RUSH

Here, in Point Reyes, herbal eucalyptus in the dry air.

What if, instead of knocking on wood every time I say Felipe and I are going to try to have a family, what if I felt rich with possibility instead? What if it isn't just my age that makes me uncertain we will be able to conceive? What if it is also because I purposefully dwell in our world's dwindling?

Today on the trail, the feeling of loss did not live inside my body. Instead my body was just a body walking down a path. The fog worked its way toward me. It came in sheets of cool, jeweled air.

MILE ～～～～～～～～～～～～～～～～～～～～～～～～～～～～～ **98.9**

Literary pilgrimages braid word and world. I start the day with a Robert Hass poem, the one that opens this way, "Tomales Bay is flat blue in the Indian summer heat." Hours later I sit on a log at the water's edge as the place itself unfurls. Wide-winged predators, tidal channels, and the bay beyond. Tule and verbena and the sound of blue thistles touching in the dry heat. Back in the marsh, I found a deer hoof—just the hoof—skin still on the shin. The shin snapped clean right above the ankle. There were flies in the air though I did not see the body anywhere.

The soil here is damp and dark. I sit surrounded by a sea of swollen-fingered succulents. People call this plant "pickleweed" because, well, because it looks like a bunch of pickles got together and formed a forest of bonsai pickle trees. Soon, I think, this place will slip permanently beneath the bay's glossy surface and seals will swim overhead.

MILE ～～～～～～～～～～～～～～～～～～～～～～～～～～～～ **102.5**

Today Black Mountain is on fire. The dark edge of the blaze looks like one long blood clot. People pull over, exit their cars. The police have cut us off at the pass. Binoculars out. Phones out. No one says anything, then back into the Prius and the pickup truck and the rented Volkswagen Beetle. We turn around and try to get to Petaluma another way.

My mother called last night. She didn't want to know about the book tour or my walks. "How close are you to the fires?" she asked. "Don't worry," I had told her, "I am nowhere near the burning." At the time it was true. But now, in my rearview, fingers of fire flare.

MILE 147.5

Elise and I are four miles into a five-mile walk when a ranger tells us we need to evacuate. Suddenly, helicopters. Dangling beneath each one a bladder heavy with water. Hot blasts of air hit my body. We are almost running. A plane flies low, pours a pink powder onto the blaze.

When we tell Elise's mother, she declares, her voice a-warble, "I just can't take this anymore." We are standing in her kitchen on Redwood Road. The kitchen that she evacuated the previous year when the wildfires were so close. I walk half a mile north to see what remains: three foundations, a chimney, redwoods with soot-scarred bark.

That night, after reading at the local public library, residents tell me they feel alone with their fear. They wonder whether they should stay or go, whether a fire will engulf their home or their family. I am learning this is how water moves, remaking what it does and does not touch. Too much water or too little. Both are equally unsettling.

When we get home, we pour ourselves glasses of white wine, easing the desire to run. During the blue hour on the edge of night bats congregate. Beyond the vineyard coyotes talk about their kill. Tomorrow morning we will eat the eggs the hens laid. Tomorrow we will stay close; we will try not to wander beyond rescue distance.

ELIZABETH RUSH

The smoke stays with me through Mendocino, and up into Oregon, through Grants Pass and Cave Junction. Because of it Crater Lake National Park is mostly empty. At the trailhead there is a chart: when you cannot see more than five miles, the air quality is bad; when you cannot see more than three miles, the air quality is dangerous; when you cannot see more than one mile, you need to leave. I squint across open space. The far cliff walls are almost indistinguishable from the sky. My map says 6.2 miles separate here from there.

Wander down from the rim of the volcano to the water's edge, then dive off a rock into the lake. Beneath the surface I open my eyes to piercing cobalt. It is the first time in over a week I can see clearly. The dark lava scree falls away precipitously until all is blue on blue, its brightness bracing. The color so close to perfect it helps me forgive a hurt I carried here and could not before let go.

ATLAS WITH SHIFTING EDGES

My friends in Portland and I drive out the Gorge to Rooster Rock. Sunbathe with infants on a sandbar in the middle of the Columbia River. A couple walks by wearing only wide-brimmed hats. The air is thick and tastes of prayer papers burning. Tommy tells me that last summer was worse. On the Fourth of July, the woods surrounding Eagle Creek and its seven waterfalls were on fire. The trail, just north of here, has yet to reopen. He tells me that Cougar Hot Springs is burning now.

The precipitation maps are contracting. The line between rain shadow and drought shifts uneasily like a tired body in a tired chair. Together we watch the water's edge wobble. We wonder what to call the feeling of losing the places that shaped us, a word for the way our very lives drive them farther and farther into the past. We eat strawberries, lean baptismal into the silty river, then click the babies into their dirty carseats and head home.

My father asks, "Is that smoke or fog?"

"Smoke," I say.

After lunch we ride the ferry to Coupeville and can barely see beyond the boat's broad nose.

My father and I came to these mountains once before, fourteen years ago. I remember the green tongues of wildflowers licking down the hillside and how stunning it was to see, for maybe the first time in my life, sharp snow-covered peaks. What we seek is not what it was. Expectation makes the heart feel full, sometimes too heavy.

Not that long ago, my father was so sick he thought he might never hike again. Today we walk the Diablo Dam trail together. We don't see much when we look out, the close peaks are blue bumps and beyond them only white. We eat Gala apples and licorice flown in from Australia. He notices my gray hairs. Above us electricity sounds twitchy in the wires. Beyond the wires the sun is a nectarine in a sea of soot. Still we are a father and daughter together in a forest.

Our campsite has no showers, only the river and its riprap, which those who stayed here before us used to make a little dam. They stacked stones to hold the water momentarily in place so that they, like Arthur, might bathe. In the middle of the night I wake to the rushing river running over those stones—tuck into the sound of glaciers melting—and fall immediately back to sleep. In the morning the smoke has cleared and the mountains at last come into focus. There, the specificity of each tree is startling.

Transfor-Mar

SAMUEL GREGOIRE

Mi delirio es para transfor-Mar,
Desquiciarme como las siete mil olas
Que llevaron Wangolo a Ziltik.

Cric . . .
Crac . . .
Tim Tim
Bwa sèch

Es un cuento de espumas saladas,
De carcajadas de siglos pasados
reanimadas en los reflejos mojados,
De silencios en vaivenes eternos.

Es un cuento de agonías perfectas
Del viento crepuscular
Perforando la esencia del horizonte.

Cric . . .
Crac . . .
Tim Tim
Bwa sèch

Érase una vez las rocas del mar
Que no sabían el dolor de las rocas quemadas . . .

Érase una vez Agwé, un poder caprichoso del mar
Vertebrando las aventuras misteriosas . . .

Érase una vez Wangolo,
Una esperanza inmadura
Condenada a Naufragar eternamente . . .

Érase una vez los "Lobe,"
Esos tam-tam violentos del tambor de mar
Tocado por Olókún,
Abundando vidas en Yemaya . . .

Érase una vez islas que flotaban
Sobre el mar embravecido
Por el esquizofrénico libido
Entre la luna y el sol . . .

No hay manera de terminar el cuento
Pero se que Ziltik es un espejismo

Así, Mi delirio siempre será para transfor-Mar,
Lamentando la maldición de Wangolo
Y llorar frente al mar
Para sentir que mis penas son pocas.

Transformocean

SAMUEL GREGOIRE, TRANSLATION BY FORREST GANDER

My delirium is a transformocean
Rocking me like the seven thousand waves
That brought Wangolo to Ziltik.

Cric . . .
Crack . . .
Tim Tim
Bwa sèch

It's a tale of salty foam,
Of centuries-old laughter
Breaking out anew in wet reflections,

Of the coming and going of silences.
It's a tale of perfect agonies,
Of crepuscular wind
Perforating the horizon's pith.

Cric . . .
Crack . . .
Tim Tim
Bwa sèch

One time it was the rocks of the sea
That didn't fathom the pain of scorched rocks . . .

One time it was Agwé, the sea's capricious power
Mobilizing mysterious adventures . . .

One time it was Wangolo,
A green hope
Condemned to go down at sea eternally . . .

One time it was the "Lobe,"
Those violent tam-tams of the sea drum
Touched by Olókún,
Full lives in Yemaya . . .

One time there were islands floating
Across the rough sea
Caught in the schizophrenic libido
Between the moon and the sun . . .

There is no way to end the story
But I know Ziltik is a mirage

So my delirium will always be a transformocean
As I lament Wangalo's curse
And I weep at the sea's edge
So that my sorrows might release me.

Wangolo is a character in Haitian folklore.
Ziltic is a Creole deformation of "Les îles Turques-et-Caïques" (the Turks and Caicos Islands).
The refrain "*Cric . . . Crac . . . Bwa sèch*" is the classic manner of beginning a traditional
    Haitian story.
"Lobe" is the action and sound made by striking the surface of water with the hand.

TRANSFORMOCEAN

Dipping In

Lenore Manderson

When you grow up on the land, you learn that water is a resource and a tool, not a plaything. It's scarce, to be safeguarded, used wisely, conserved.

≈

Rain was hardly an issue on our first block, a dairy farm some fifty miles south of Melbourne. Set at about the lowest point of the district, boiling water to warm the bath must have taken forever. It rained persistently, and when it wasn't raining, it was misty; rain was always in the offing. The back paddocks were sodden; cows were routinely sucked into and stuck in the waterlogged soil. We pulled our legs out of our gumboots as we dragged ourselves free, and messed the kitchen floor with clumps of mud on our return to the house to get dry.

Yet I have no recollection of water glistening on leaves, or of raindrops on the tin roof. Instead, reinforced by faded photographs and family story-telling, I hold much drier memories of my childhood in the fifties. Water was an entertainment. In late spring, my two younger brothers and I would cycle down to a pond rippling with tadpoles and dragonfly larvae, framed by bracken, flowering egg-and-bacon plants, and a litter of dried leaves and twigs. In summer, we'd play with our cousins by the bore water tank set between the cow shed and the house; the water smelt slightly fecal, slightly metallic, living. On the hottest days, as the sun slipped and softened, we'd run in and out of a sprinkler's tepid spray, wearing only underpants and already self-conscious about it at age seven or eight.

≈

Our father was a strong swimmer. Unlikely as it seemed to us (but we verified it), he won a six-mile race down the Yarra River through the heart of Melbourne, doing the breaststroke, soon after he migrated. We never asked where he'd learned to swim. He certainly had no options at the orphanage in Johannesburg, nor the north African desertscape of war, nor the camps of Italy and Germany. Maybe he swam in a nearby pool as a school sport activity or as weekend recreation, a break from the boredom of orphanage confinement. In any case, he swam strongly. He took pleasure in the water; on it, in boats, he just got sick.

Our farm was about eight miles from the nearest beaches, with names of local derivation like Jam Jerrup and Lang Lang. These beaches were for paddling and fishing, their shorelines more mangrove swamp and mud than sand.

So we traveled further, twenty miles to Tooradin, but rarely—time not distance shapes the possibility of leisure for farmers and their children. The town's name derives from the local word for "swamp monster" or "bunyip," given to it when it was claimed by white colonists in 1840. In the 1950s, it still had monster spookiness, heightened by the island prison farm in the middle of the bay. It's a place now touted as a secret paradise of serenity and stunning views. We imagined violent recalcitrant prisoners, plotting their escape to the mainland. Our grim view of the island and the bay was bolstered by the mangrove trees, with their mysterious roots and gray mudflats, and by the seaweed that would wrap around our legs. Here, we learned to float and swim in the shallows. Afterward, our father stretched out in the brackish water, swimming back and forth, one eye on us as we built sandcastles and moats and grew sweaty and hungry.

Then he'd lead us through his repertoire of war and music hall songs as we bounced our way home.

We're on the road to anywhere / With never a heartache and never a care.

~

In 1960, we left this farm because of its wetness, the unrelenting damp-ness of the place, and because of the hard routine, the milking at 5 a.m. and 5 p.m. to maximize yield. We moved to another farm, wheat and sheep, until our father decided that sheep were even more stupid than cows, and he converted it to dairy too. We were old enough to hold the pressure hose to the concrete after milking, adding water to a fetid mix of urine, dung, and milk slop.

The farm was two hundred miles north of Melbourne, a soldier-settler block nested between the two longest rivers in the country, the Murray and the Murrumbidgee. By surveying accident rather than design, we assume, the farm was too large for one farmer, too small to be share-cropped. It rained coldly in winter. Summers were dry, although the front paddock was always oddly swampy, a hide for snakes and a quarantine for bulls and feral goats. And because the summers were dry, the land was irrigated, the channels tracking the fencing. We three children, in shorts and gum boots, our eyes scanning for snakes and rabbits, would follow our father as he cut into the banks to flood the paddocks.

The year takes its rhythms from the water and its seasons. Our father, new to Australia and to working the land, learned that the key to neighborli-ness was acknowledging this. *What's the weather going to do?* he'd ask the neighbor he'd ring each morning once we had a phone. I'd listen in and down the other end of the line, Bill—the nearest neighbor was Bill at both our farms—might reflect on the possibility of rain, though tersely. *Might rain.* The call was over in a moment.

We needed the rain to stay on the land. Rain filled the corrugated iron tanks by the house, providing us water for cooking and drinking. The dams outside the garden gate provided water for laundry and bath water and drinking water for the animals. Enough rain, a full tank and a full dam, and there was water for the garden too.

I thought his phone call was uniquely funny, but it wasn't. Farmers always asked about the weather when they met in town or passed by on the road. Now there is a national website, farmonlineweather.com.au. "Medium chance of showers, most likely early evening," "The chance of a thunderstorm," "Slight chance of a shower." While the web forecasts threaten to steal from farmers their few opportunities to socialize, the nuances of chance, typology, and volume of rain must surely allow for a good yarn.

〰

The nearest swimming hole was the house dam, a place in which we could cool off at the end of the day, always watchful for snakes and sticks, the bull in the neighboring paddock, the rare corpse of an unlucky calf. The dam watered the orchard and the vegetable patch, fed the laundry and the house; the water splashed rust-colored out of the bathroom taps.

Occasionally, we'd bundle into the car and go to the deepest, widest channel in the far corner of our farm. We'd wade in timidly, as gray mud squelched between our toes, wary of nips from yabbies too small to catch and eat, still more wary of leeches as they'd slither toward us. On rarer occasions, between milkings and before the sun was too high, or when late afternoon summer heat cut short farm work, we'd head for the large irrigation canal that fed water to the channels. We'd all be there—our mother and baby brother too—finding respite from the suffocating house. Cows would wander along the opposite bank, or watch us from the safety of the other side of a fence. We'd swim more and play less here than in the dam or the channel; the canal's steady current kept us on the move. On the way back home, sweaty in the back seat of the car, we'd continue our ritual end of the day, more excerpts from our father's songbook. *Wheear 'as ta bin sin ah saw thee? / On Ilkla Moor baht 'at?*

Less frequently, because it was a day away from the cows, when there was no risk of a thunderstorm and lightning striking wheat stubble, we'd drive to the Murray. When our baby brother was born, and over the days he and my mother rested in hospital, as was practice then, we'd peek in to admire him and hug her shyly. Then, excitement high, we'd cross the bridge and drive along other lower bridges that spanned the backwaters, broken up with snags and woody debris, proof of river floods. We'd spill out of the car with towels, buckets and spades, and fruit. We'd head for the sandy beach and into the bronze water, chased by warnings of hidden stumps and quick currents, and cautionary tales of diving injuries and drownings. The only near-drown we saw was of a dog, not ours, but our father saved him. Witnessing this feat, we gave him the central place of heroism in the annals of our childhood.

〰

Our mother dreamed of beach holidays and a beach house; these were powerful counterpoints to the working land's boundaries. In our childhood, the farms left us cash poor, and so we stayed off-season in modest cottages, spending our time bundled up against sea breezes and rain as we walked along the waterline. We looked for shells of beauty, sponge, cuttlefish bone, broken bits of jellyfish, a dead fish. We'd pop the bubbles of the beaded seaweed, kick over the algae and kelp and ferns. We'd follow tiny sand crabs into holes. We'd look for fish and periwinkle and anemone in rock pools. We'd watch tough swimmers brave the cold gray water. We'd eat fish and chips in the warmth of the car while, as windshield wipers allowed, we took in the view.

LENORE MANDERSON

When we grew older and moved to a city, and as public swimming pools became commonplace, we swam there. And even the pools, clear turquoise, transparent and smelling of chloramines, came with warnings: a toddler facedown; a child in trouble too soon in the water after a heavy meal; a death from a dive at the shallow end. If lifesaving certificates hadn't been compulsory, and pools hadn't given us a chance for a little free time and to show off new bathing suits, we might have steered clear.

We didn't grow up on the coast. We didn't have years with the sea as our playground. We learned to bounce around in the shallows and to body ride occasional waves near the shore. But the seaside was not our habitual rest place. We didn't surf or sail or fish or whale watch. We didn't own a pool, but we did still cool off in the heat of summer by a river at the suburbs' edge. The water we knew best was opaque and unpredictable, its currents and riverine life and hidden rocks and stumps demanding our caution rather than abandon. We learned to respect water and its sources, and to ration its use.

Roots

COLIN CHANNER

Then, the future was glaucomic, the bore through mangrove
in the dugout slow. I recall the water in its color tannic.
I see now an olive wake dissolving from the churn work
of the screw. A time would come—it seems it has—
to redecipher, understand again the meaning of the motor's
open vowels louding up a sacred space.

Corporal Pitt, the bully, said something far beyond himself,
"You see all what favor frame for madman basket?
those are aerial roots." He pointed and we took
his reedy finger as command, us six good recruits—
cadet acolytes joined for camping life—and paused
eye-sweep for crocodiles.

I plait time to those wetlands often. To be black where
I live now is to bivouac. White is wilderness in all seasons.
I carry bankras of one-one sorrows; gods in a haversack of joy.

Out on long lug-sucking walks through marshes south of Boston,
close-west fairly of the Cape, I wink "like" to the look of bulrushes,
how they call to bible Moses, kinda favor sugarcane.
Who resists the cattails' saucery?—such flirts—but the names.

Little Massachuck. Sachuest. Sapowet.
Say them soft; no, shout these native names,
names of the plowed near, and housed to,
the made margin, the selvedged by road,

the done to as America tends to do with indigenes,
its what-it-failed-to-kills.

At water's edge a man in waders arcs a lure; snaps it
out for bass. Tammed women with clam baskets hunch
against a pushy breeze in group leverage. Seashells smaller
than the ears of newborns crunch in the wake of boots.
My dry-meniscus knees go skurch on pebble shoals.

Sinuous chapel festooned-gaudy, by ibis candle-lit,
I sight you. But how I coulda note full conscious
your low-key frieze of halophytes, the mangroves' gazing
wall of Afroed saints? I was just manyouth.

Once, I pilgrimed to another coast of my island
to be witnessed to in soulcase by the final two
uncaught unkilled sea cows,
figures so of there, but as their wakes were, evanescent.
They're gone now like the Arawaks as I too must go.

I go. Home for now to Providence. Comb-somed,
bearded, chukking old Bean boots—apparently adaptive.
Every hair a root.

Highwater Nocturne

Kulvinder Kaur Dhew

Highwater Nocturne, 2019. Charcoal on paper, 14 × 21 inches. Copyright © Kulvinder Kaur Dhew. Private collection, New York, USA.

Living with the River

COLE SWENSEN

WATER, THAT MEDIUM WE SO ABSOLUTELY CANNOT INHABIT, NONE-theless shapes so many of the places that we do inhabit. Beginning with the fact that it often selects those places.

I ran across this statistic in an online journal published by the National Institute for the Humanities: over 50 percent of the world's population lives within three kilometers of a body of freshwater, and only 10 percent of the population lives farther than ten kilometers away from one.

I was particularly interested in these statistics because I've been think-ing about rivers and about the specific kind of living arrangements that they enable and support. Among bodies of water, rivers are distinct, in that they're *running*, so they not only offer water, but also power and transportation.

I'm going to use the Gave de Pau, a river in the south of France that runs at the foot of the city of Pau at the base of the Pyrenees, as my example. A couple of years ago, I was doing research on the river that addressed three questions: How does a river determine the lives that live along it? How has that changed over time? And what do those changes tell us about other changes in economics, culture, values, and so on?

In addition to their myriad practical advantages, one of the major attractions of rivers is the metaphoric. They radiate deeply binding associations with time, the flow of life, and the passage from life to death, among others. To live beside a river is to live in constant connec-tion to and conversation with these powerful and eternal conundrums.

Another of their attractions is aesthetic. The American poet Lyn Hejinian has a line in her book *My Life*: "It's hard to turn away from moving water." This simple yet irrefutable truism points to something important in terms of aesthetics. It's not a question of being "beautiful" per se, but of being compelling, and, as she insightfully points out, it's the motion that is compelling; it's as if the various circulatory systems in our bodies—blood, lymph, mucus—all respond at a visceral level to this motion. We have a sympathetic response to it.

The aesthetic activates other senses as well; there's a strong audio component to a river, one that's constantly changing throughout the seasons and according to the context of the banks—the sound as a river goes over shallow rocks as opposed to the sound in deep places, as it rounds a graceful bend, the absolute roar of a waterfall. All these sounds, in addition to their musical component, are also constantly giving us crucial information about the river, its depth, its conditions, its speed.

And, of course, there is a strong, and for many people quite attractive, visual aspect to a river. Water itself is a metaphor for life, and whether it's clear or cloudy, its surface troubled or calm, its attitude seems to shed its mood over our day. Clear water, such as that in the Gave de Pau in late summer and fall, alters objects in the riverbed, making them appear endlessly changing, so that even the most mundane objects, such as stones, are constantly served up new. The surface of a muddy river can be just as intriguing—we focus on the surface as a surface, reading its colors, enjoying its textures. And when we say "picture a river," we don't picture just the water; a river is also its framing, its context, the banks, cliffs, or pastures that hem it in. Because of the ready availability of freshwater, these are often vibrant places, teeming with green life that bursts into flower each spring. The latent equation between beauty and health is brought out emphatically in such places.

The aesthetic, though, has a more complicated side in this case; I touch on this later. But to return for a moment to water and habitation, and particularly to the notion that water often chooses where we live: this is the case with the city of Pau, founded at this particular spot because it was a natural ford, the only place the river could be forded for many miles in either direction. Millennia earlier, the river had carved up the region in such a way as to leave a promontory with a cliff face dropping down to the river, which made it both an easily defensible location and one safe from the river's frequent flooding. A defensive structure was first erected on the promontory in the eleventh or twelfth century, protecting the ford, which was crucial for getting the herds up to the meadows of the Pyrenees in the summer months.

Over the centuries, Pau developed steadily, with more and more of its activities centered upon the river. In one sense, the river literally built the city; it was the source of the three principal building materials—stone, gravel, and sand. Most of the pre–World War II buildings in the area are built of the distinctive rounded oblong stones that are polished and washed down from the high Pyrenees by the river. This creates a startling kind of doubling: you look down into the river, which in most places and depending upon the season is shallow and clear enough to see to the bottom, and you see the countless oblong stones that line it, and then you look up and see these same stones rearranged into houses, barns, paths, fences, gateposts. . . . Sand from the river is used to make the mortar that holds them all together. Among other things, then, the river is a factory creating the materials that in turn create the city—and strongly determine its appearance. Through its abundant water, the river is also the indirect supplier of the wood readily available on its banks—both for building and for burning as a source of heat before other fuel sources were available.

The Gave de Pau is not easily navigable; it's particularly as a source of power and water that it contributed to the region's industrial development. The first industries were mills, and there were many up and down the river, often on detoured canals. The first canal, the Canal du Moulin, dates back to the early thirteenth century. These canals allowed millers to have greater control over water flow than the fluctuations of the river itself allowed. Other industries that developed included laundries, tanneries, paper mills, and textile mills. By the late eighteenth century, the region was widely known for weaving. Some 30 percent of the local artisans were weavers working in both linen and wool; they were particularly known for their fine linen handkerchiefs.

Flooding significantly influenced the life of the region. Globally, community conversations around water usually tend to be concerned with its scarcity, which plays a major role in the politics of the many arid regions of the world; here, it's an overabundance that needs to be managed, and one of the reigning questions is that of permanence vs. impermanence: Is it better to build easier, cheaper wooden bridges, knowing that they'll get washed away at every flood? Or to build hefty, expensive, labor-intensive stone bridges, hoping that they won't? Investors in factories and industries along the banks had to take the dangers of flooding into account, while farmers, though their crops might be devastated, also benefited from the new deposits of topsoil, and herding peoples, the most flexible and mobile of the three groups, provided they could get their herds to high ground in time, suffered much less.

Today, in part because of this uncertainty, there are relatively few buildings right on the banks. The industrial face of the place has entirely changed—there are no more mills, laundries, tanneries, chocolate factories, or paper manufacturers. The larger reason for this is the simple fact that they don't need to be there because they don't need to get their energy from the river—such industries now get it through less direct means. And thus, the disappearance of these riverbank industries says a lot about our increasing distance from our sources—in this case, from the energy we use to produce things. The increasing availability of electric and gas power frees us from the necessity of specific locations. But this means, in turn, that people in the region who were once in constant contact with the direct source of their livelihoods, and with all of their senses—sight, hearing, smell, touch—now have a much more abstract relationship to that which actually drives their lives. It also meant that they were giving up a free and renewable source of energy for nonrenewables that cost both them and the planet considerably more.

In just a couple of generations, the river's function has completely altered; it's shifted from its role, front and center, as an essential collaborator in daily life, to the margins, where it's become simply a pleasant background, whose principal use for most locals is as a site for leisure activities. For a mile or so directly west of Pau, the north bank of the river has been transformed into a long park with grassy stretches and attractive groves of trees. There's a golf course and a walking path that continues on for some fifteen miles. Signs posted at irregular intervals show the paths and their mileage and mark the locations of various other amenities—a lake with café and picnic grounds, riding stables, bicycle rentals. . . . A thriving kayaking, canoeing, and rafting culture has developed on the river itself, with guided trips for tourists and classes for locals. The river is also still used for fishing, not fishing as a life-sustaining source of food, but as a relaxing activity, a Saturday morning diversion to contrast with the stresses of the working week. And so, the river drifts in the popular mind more and more away from the very basis of the local economy, the key enabler of daily life, toward a mere aspect of décor, which, in turn, changes the population's sense of responsibility toward it.

The focus is now on keeping it looking attractive in a postcard kind of way, which means that the aesthetic questions around it are no longer part of a broad and complex understanding of environmental health, of keeping many elements in balance so that herds, agriculture, industry, and fishing all can thrive. Instead, care of the river is reduced to a narrow concept of "beauty" based upon its superficial appearance, which is, in turn, based on preconceived and imported notions of parks and gardens.

And while I think this situation warrants deep consideration, I'm not saying that it's *wrong*, and I don't want to idealize the past, in which disputes over the use of the river were rife. It's appropriate that things change according to circumstances, even things as big as rivers, and environmental issues are taken seriously today in ways that they never were before. However, in this case, the narrowing of aesthetics to cover only a certain kind of visual beauty is actually masking problems that are endangering the health of the river.

And here, we must stop to think out that phrase—when we talk about the health of the river, what do we really mean? Water cannot in itself be healthy or not—water, or bodies of water, can be *healthful* or not, but that's a different thing and entirely subject to the particularities of the organisms engaging with it. Water that includes certain chemicals, for instance, might be quite unhealthful for mammals, while offering the ideal medium for certain algae to thrive. So, when we think about the health of a river, we're thinking about more than the water; we're thinking about a complex ecosystem of plants and animals for which the body of water (the river, in this case) is a, or perhaps the, common denominator. And we're thinking about it from a human perspective—that which helps us and our friends thrive (friends defined as cattle, sheep, fish, rabbits—we may kill them all for our own gain, but we find it tragic when they die of drinking "polluted" water). A different set of organisms, bacteria, for instance, would define pollution very differently and thus have a different list of priorities.

So, to talk about the "health" of an ecosystem, we must accept that we're talking about it with human bias, not unquestionably, but understandably, in terms of its healthfulness for us. Starting from that acknowledgement, what constitutes health? Or, phrased differently, "What are we trying to achieve?" When we try to "clean up a river," what results are we after? One definition of a healthy ecosystem might privilege diversity, and consider that the highest number of different species of plants and animals (that don't harm us), the richer, and therefore the healthier, the environment.

Back to the example of the Gave de Pau, the current population of the area tends to assume that this is the case, that diversity indicates health. The riverbanks are thriving with trees and vines and other green plants, and, with its now more distanced relationship to the river itself, based almost exclusively on sight, most of the population assumes that, since the water in the river looks very clear, it must be "clean"—though what do we mean by that? Do we mean "composed of very little other than H_2O"? But in fact, the river is more polluted now than ever, if pollution is taken to mean "that which might harm humans and our friends." And that is saying a lot when you think of the large-scale polluters, such as the tanneries, laundries, and other industries that have moved.

But it is highly polluted, and apparently, it's just a matter of numbers—there are so many more people living along it than ever before in its history, and it has so far proved impossible to keep all of the sewage out of the river. A water treatment plant has been built at Lescar, just down the river from the city of Pau, and a major study was done in 2012 to address the issue, but the sewers of the many cities along the river still share their canalization with the runoff from the streets and gutters, and whenever there is especially heavy rainfall, or even the slightest flood, used water all goes directly into the river.

Inveterate kayakers and canoers—those who experience the river through other senses in addition to sight, such as touch and smell—notice the pollution, as it causes irritation to the skin, and so they are campaigning for change, and various cities along the river have engineers looking into the problem, but few others in the general population are aware of it, and so it hasn't become a top priority.

This situation offers an example of a population operating from a narrow notion of aesthetics, one focused on the single sense of sight. It's narrowed even further in much of Europe and North America by our inherited sense of beauty based on a classical conception of an ideal landscape such as that promulgated by the seventeenth-century painters Poussin, Lorrain, and others. Movements such as Impressionism and much of Postimpressionism, which in other ways opened up aesthetic possibility, in this one sense, reinforced the narrow view. And this narrow view prevents our seeing beyond its terms, other than through the term "ugly." It has created a binary that blinds us to complexities that may fall between the embedded categories of the "beautiful" and the "ugly" and yet fully embody the "healthy" and the "healthful."

One brief example: Gilles Clément, a noted French landscape theorist, has developed the idea of "the third landscape." This is the seemingly abandoned zones—empty lots, highway shoulders, the banks of drainage ditches—between agriculture and other sorts of intentional plantings. Through benign neglect, these zones foster an incredible diversity of plant life in regions that otherwise restrict it through herbicides, weed-whacking, and other activities geared toward keeping the landscape as a whole looking tidy and well cared for. It's an excellent example of an understanding of beauty that has been reduced to the point that it works against, not for, a heathier and more healthful environment.

Still, Life

AILSA PIPER

I'M NOT AWARE THE RAIN IS FALLING UNTIL I SURFACE. DROPLETS HIT the water then bounce up again, defying gravity, backlit against dawn light. For an instant, harbor and heavens connect via thin strands of liquid—the droplets rebounding, shimmying one last time before finally dissolving into the harbor, dancers no more. They've become part of the whole, caught in the heave of the tide. Like me, they're absorbed in this warm, velvety sway.

I flip onto my back, letting the force of the summer downpour splash onto my goggles. Drops thunder in my ears, in just the way my racing heartbeat blocked out all other sounds when I was first learning to swim, just three years ago. Now, I'm barely aware of it, and when I check my heartbeat, it's even and steady.

I've passed another test.

Buoyed by salt, I float, a still thing between universes of movement. Clouds bump and merge, swirling steel gray through dove gray through pearl gray. This day will be soft.

I flip again and begin to stroke toward the pitted sandstone headland. It too is gray this morning. Most days, I swim between the gold of the rising sun and the burning umber of those rocks. Today, as I stroke and count, the world is the color of a tern's wing, like the one I glimpse as my head rotates to breathe. Left, two, three. Right, two, three. Left, two, three . . .

I recall the voice of my coach.

"Focus on the exhale," Marina says. "Keep your breathing easy."

Marina is a mermaid, a childhood champion freestyler. She's an artist who works in oils, inks, and gouaches, capturing urban rooftops and gardens; portraits too, though less often. She also paints flowers in vases. Still lifes, they're called—not still lives. That's another thing I've learned from keeping company with her, that correct-but-confusing plural. Still lifes.

I've been picturing one as I swim . . .

Six months ago, it traveled all the way south from Amsterdam's Rijksmuseum to Sydney—16,636 kilometers. I took a fifteen-minute ferry ride to the art gallery, paid my entry fee, wandered past the other visitors (Rembrandts and a Vermeer among them), and turned a corner into the room of still lifes. There it was. It had been waiting for me to find it for more than three hundred years. It is as clear today, in my swimming goggles, as it was on that first sighting . . .

Six skulls lie at random angles, discarded, on a stone benchtop. They are yellowing, their foreheads and cheekbones warm gold, save for one bleached specimen that sits upright, front and center, its gaping eye sockets staring out into the gallery, brazen in its unflinching disdain. It has a high forehead, this cranium. There had been a big brain inside it; a big mind. What had stopped it? Behind it, a darker-toned skull reveals its interior—the cavities for teeth, the ridges of the roof of the mouth, the gaping hole for the windpipe. No sound will ever be made there again.

Who were these six skulls, when they had flesh covering them and blood coursing through them? How did they come to be a pile of remnants, abandoned with other bones in this dark corner?

A discarded lower jaw sits alone at a distance from the skulls, many of its teeth missing. Behind it stands a wooden hourglass. The sand has run out. A single candle is the painting's sole source of illumination, and soon that will be gone—the flame has almost reached the socket of the candlestick. Time is limited for the painter.

On a shelf behind the skulls, documents battle for space with cloth- and leather-bound books. These volumes were used—they are bookmarked and dog-eared. Maybe, given the two red seals dangling from the shelf, they were a lawyer's reference books. Perhaps the folded papers are last wills and testaments.

The skulls grin. Important papers are nothing in the face of mortality.

Behind the hourglass is a pedestal of sandstone or marble. The artist's name—A. van der Schoor—is carved into it in blood-colored script. What was his mental state as he painted this image? To ponder such objects for days or weeks would be to flirt with internal darkness, surely?

In the foreground, at the edge of the bench, lie two summer blooms, like those my mother described in the letter she wrote to me when she was dying.

Whenever you see pink roses, you will know that I am near.

I couldn't feel her at the gallery.

Giggling schoolchildren crowded around me, cracking jokes about the fate of the owners of the skulls. I was jostled close, too close, to the bones and the dying flame of the candle. My chest constricted. My heart began to pound. Heat rose up my spine and into my scalp. I shut my eyes and focused on my breath until the students lost interest, moving on to the next picture like a flock of predatory birds.

Even with my eyes closed and the hubbub receding, I couldn't lose sight of those blank eye sockets and the dark nothingness beyond them. When I

STILL, LIFE

opened my eyes and moved away, the image still hovered about, just as it has been floating in the water this morning. Surely those are pink petals drifting above me, beside me . . .

One thirty-two, one thirty-five, one thirty-eight . . .

Count the breaths. One forty-one. Keep counting. Breathe and count. One forty-four . . .

It's the only way to bring myself back when anxiety threatens—and mostly, it works. One forty-seven, one fifty.

I've not had the panic for weeks, but here it is again, that unwelcome visitor.

I woke in the night because I had dreamed. I came to in darkness, surfacing like a swimmer sucking in air after diving too deep. I sat up, turned on the light, and went over the images, recounting them to myself so I could hold the feeling, keep the dream. I told myself the story of it, as though reciting a fairy tale to a child.

> There was a group of us and we were sad.
>
> I was very sad and worried.
>
> Then Peter arrived. My husband. Alive!
>
> We were safe.
>
> I felt happy. We all felt very happy.
>
> I ran across the room and threw my arms around his neck and the others laughed and he swung me around and everything was all right. He said that as he swung me around.
>
> Everything is all right.

I couldn't stay awake, traitor that I am. I couldn't stay with him. I went back to sleep, snuggling into the safety of being held. Everything all right.

And in the oblivion of sleep, I lost the dream.

Is that why I didn't sleep after he died?

I was frightened that I would lose the story?

The panic attacks came then. I had never known them in all my life, but they were my new companions.

But this morning at dawn, after the dream, panic lost the battle. Swimming has taught me something about breathing and controlling. I sat up in bed and counted in brackets of three. Suck in the air, exhale for three. Suck in, exhale for three. By the time I arrived at the water, the dream still lapped at my edges, but I threw myself into the harbor. I stroked and counted, breathed and counted, until a cavalcade of numbers replaced the stabs of loss.

It's not possible to cry when freestyling, I've learned.

Saltwater washed away the dream, and the remembered still life, so that eventually I was able to see whiting skimming the harbor floor. They're only visible in motion, their color such a perfect match for the sand that they disappear when still.

Now, little toadfish flick their gills as schools of skate flit above them. Then there's me, stroking along the surface, raindrops dancing on my back. As my head rotates left, I glimpse others on the shore, scurrying to rescue beach bags from the downpour, while to my right, beyond the net, a multistory cruise liner makes her way toward port. Just ahead, a creamy skate ruffles its fluted edges and is gone. My hands cleave through a school of darting silvery streaks, tiny fish whose name I don't yet know. There is so much movement in this underworld, so much flickering, darting life, even in the shallows.

When I was learning, I had to stare down the deep—heart pounding, lungs pumping—as I swam against fear. Of sharks, yes, and stingrays, too. But the real terror was of other, bigger, things. Worse than any sea monster's fangs was the dark shadow that I knew could swoop at any moment, to steal, to still, life.

I rotate my shoulder and extend my arm. My eyes scan the harbor floor as I exhale through three more strokes and rotate again.

On the shore, I spot Esme keeping dry under the kiosk's canopy. She'll turn ninety-nine next Saturday. There's to be a party here by the water. We swimmers are her family, she says. On my next rotation to the left, I sight her caregiver, a slight girl in thin cotton sweatpants, lugging a black bucket across the sand and up the concrete steps to where Esme waits.

Rotating to the right, I glimpse the raindrops. They are diamonds on the surface.

Next rotation to the left, I see Esme lean forward to dip her head into the bucket of brine. Kiosk patrons will be concerned, just as I was when I first saw this ritual. Esme has cancer. It won't kill her, she says, but there is no repair for it either. It's a slow-moving form, and she can't predict where it will next appear. For the last year, it has ravaged her skin and she has not been able to swim. She has open wounds on her legs—a gash of flesh bleeds constantly. On her face are scabs.

She misses the saltwater, so her caregiver brings it to her. Sometimes, the bucket is for her legs—she lowers a shin into it and stands, twitching at the bite of salt on the wound, before swapping sides. Other times, the bucket is placed between the handles of her walking frame, so she can plunge her face in, holding her breath in the silky, salty darkness. When she emerges, her thin gray hair will be dripping and the scabs will be soft.

Esme says old age is lonely. She remarked, two days ago, that the beach is populated with widows, abandoned by the world because they are partnerless. I've never told her I'm a widow, but now I look at the other women in their black costumes with new eyes. It hadn't occurred to me that our swimming togs were funereal.

"But we keep coming," Esme said, pulling hair back from her face and licking the salt. "Down we all come, to the lonely sea and the sky."

I am profoundly grateful that grief led me to this water. Daily, I observe those who've lived long in the curing properties of salt. I see their knots, their lumps, and their creaks, and I hope that maybe, just maybe, I too might outrun a few more years if I can keep navigating between the sandstone rocks at either end of this sacred curve. If I'm lucky, I will get to have more age spots and wrinkles. If I'm really lucky, I'll be the one generating awed whispers that anyone could be so old and still be swimming, even with her crow's feet and corrugations, her crepe skin and rheumy eyes.

I breathe. I stroke. I swim.

The Waters Will Remember

Atul Bhalla

The waters
remember
shore to shore
stain to stain
lost
and found
the waters
journeys
I know
that you knew
the smoked-out dreams
labor denied
exiled
memories
witnessed
on the shores
Sabarmati
the waters
will remember
what we covet
what we apprehend
truth
was
and belonged
the waters
will remember

Nostalgia IV, 2019. Archival pigment paint, 20 × 45 inches. Copyright © Atul Bhalla.

Gulf

Kulvinder Kaur Dhew

Gulf, 2011. Charcoal on paper, 38 × 50 inches. Copyright © Kulvinder Kaur Dhew. Hoffmann Collection, New Orleans, USA.

Sourcing the Stream

Wendy Woodson

APRIL 2017. IT'S SPRING, THREE MONTHS AFTER THE INAUGURATION OF Donald Trump. The mood is still dark and somber, even with the coming of new light. How to find some reassurance that the world will go on, despite the pervasive feeling of despair? Bella, the dog, signals it's time for a walk. I'm tired of the usual paths. So I take her to water, to the stream. A long winding dirt path laced with thick entangled roots and winter's debris. We walk for hours at the water's edge. The rhythm of the walk and Bella's delight change the mood, and the voices in my head begin to sing in a different tune. Stream song. I decide to spend each day walking with the stream, looking and listening for solace.

A diary, after the fact.

~~~

It's a beautiful day of blue and clouds and beginning spring leaves reflected in mesmerizing patterns in the stream. I stand at the riverbank waiting to be carried, carried in my mind, stream mind moving downstream, rapid water and detritus, all toward an inevitable destination, I think ocean. Singing in repeated refrain. Today I collect two hours of stream motions and designs on my cell phone. By now I have a total of ten hours of the stream's patterns, reflections, and abstractions. It is a kaleidoscope, an ever-shifting canvas, different each day. It has become a visceral necessity for me to observe these daily changes in the stream and its terrain. I am grateful that I have this place for walking in step with the movement of water and with my constant companion. Bella seizes every opportunity to leap wildly into the water—her play is an integral part of the love I feel with the stream.

~

Today rain changes the stream. I hear chanting falls of repetitive rhythms and watch the raindrops create echo circles on the stream's surface. I follow increased flow and then come to sudden logjams, snarls. How to get unstuck, out of a jam? The small piles of leaves and twigs caught in rock crevices in the stream wait patiently for new water from the force of rain to carry them downstream. I take this as a lesson to remember, another lesson from the stream. I start to imagine many different ways a human body, or group of bodies, can be carried down a "virtual" stream in performance. I go to other characteristics of water as inspiration for choreographic invention: thermal capacity, density, specific weight, surface tension, conductivity, boiling and freezing points, viscosity and cohesion, attraction to polar molecules. Water behaves like a magnet with an electric charge, two hydrogen and one oxygen; water is transparent, takes heat in, radiates heat out slowly. These are all powerful jumping off points for dance invention; the stream has everything we need.

> Freefall fast and
> furious
> heartstring
> plucked and
> played in river
> song well-worn
> stones and rush
> keep coming

WENDY WOODSON

the rain falls and flows
downstream glint and
glamour coming
undone

≈

I pay attention to the stream of conversations and child's play as I walk
along the path. I catch snatches of dialogue. "So, what I really want
to know is where did all this come from?" "If only I had known then
what I know now." "Did he ever get back to you?" "Yeah, it's hard to
know how to draw the line." "When do you have to let them know?" I
watch a handful of children laughing and screaming in delight as they
throw stones across the stream or build little stone castles and watch
a current knock them down. I enjoy these knock downs, I want to see
similar blows to some of our political problems. Dogs bark across the
different banks; birds call from one location to another. The place is
filled with miraculous orchestration and intricate systems of collaboration.

≈

A strong wind whips the stream, shifting the way things cooper-
ate, the way they move with or against each other. Many new images
come hurtling into my mind: surge, storm, unrelenting, gray, bending
almost broken, needing relief, what can we depend on, radical unex-
pected changes in landscape. Yet I am excited, even in this kind of wind
despair, to get new images of the stream in my cell phone. I try not to
question my drive as an artist—to capture and use everything I can to
make something at almost all costs. I hold on to the "almost." I won't
do *everything* to get the right picture or moment. In holding on to that

moderation, I put my cell phone away, into my back pocket. I simply
walk in the wind and absorb. We will be back tomorrow after the wind
dies down.

≋

I am wanting to know some
truth.

I can hear myself thinking
I want these truths to be like
feathers falling from the
sky in flight or in flow.
Celestial songs from the
deep temporary timbres
speaking in tongues.

I am curved into a comma trying to
lie between phrases the possibility
of bridge and some kind of
crossing. I am waiting long and
attentive no words or constructed
metaphors. Only song, water over
rock occasional bird.

≋

Slowly you take in your surroundings.
There is someone standing next to you
looking out at the same story the story
of sound perhaps sand. You are both

WENDY WOODSON

new to this situation a different
occasion for getting to know someone else.

Will it be through language,
words imagined but unspoken.
Or a subtle gesture, turning of
the head looking out at the
moving water light and
shadow stream.

〜

July. I have accumulated word and video images of stream in every kind
of day from spring to summer. I have been invited for a week residency
at the A.P.E. Gallery in Northampton to use the space in whatever way
I wish. I want to use the stream as subject. I want to try and create a
structure and way of directing a cast of nine dancers with lessons I have
learned from these walks. Follow the flow, play with ever-changing patterns and reflections, symphonies of sound and signals of collaboration, patience and solace.

Eurycea

Zoe Nyssa

Eurycea!

I'm here.

I can't see anything.

I'm right next to you. Spread your fingers.

There's no air. I can't get enough air!

Stop thrashing! You need to relax. Just tilt your head. There. See?

Are you sure it's enough?

What, the cavities? You'll be fine.

It's not like I imagined. I think somehow I didn't expect it to be so dark.

What would you see anyway? The currents tell you everything.

You're being cryptic.

I'm not!

No, I mean literally. Here:

> *Eurycea waterlooensis* is what biologists call a "cryptic species." A salamander measuring just an inch or two long, with an indistinct, unpigmented appearance and flattened snout and body, *waterlooensis* looks rather larval even in adulthood. Failing to grow functional eyes or lungs and never undergoing metamorphosis, these species spend their entire lives underwater in subterranean cavities, perpetually cool despite living in an arid climate. Small, wet, and nearly faceless:

unsurprisingly, many find these animals downright ugly. If environmental conservation often focuses on so-called charismatic species and landscapes—cuddly pandas, proud bald eagles, majestic giant sequoias, and breathtaking Grand Canyons—surely *waterlooensis* represents the anti-pandas of the world. Preferring to keep to underground aquatic depths, they are difficult to observe at the best of times and even experienced biologists can find *waterlooensis*, and its sighted near-cousin, *Eurycea sosorum*, hard to identify.

Who wrote that?

I did. In my PhD thesis.

Indeed, though the species shares a habitat that has been among the most intensively studied in the world by herpetologists for over a century, *sosorum* was not formally described until 1993, its existence until then considered of too little interest to be more than sparsely documented in biologists' field notes and reports for much of the twentieth century. Meanwhile, despite its notable lack of sight organs, the blind *waterlooensis* was not identified as a distinct species until 2001, its discovery an accidental byproduct of scientists looking for *sosorum*. Genetically, it is unclear exactly how the two species fit into the larger evolutionary family tree, nor is it certain just how many of these two species of *Eurycea* there are, their lifespan in the wild, how many young they produce, what they prefer to eat, or are eaten by.

What do you think?

Did you get your PhD?

Yes.

Well, then what does it matter what I think?

You don't like it.

I don't love it. "Failing to grow eyes"? "Faceless"?

I was trying to do justice to how extraordinary you are.

Even that most basic fact of wildlife guides—the area where the salamanders live—is mostly a matter of guesswork as the extent of the animals' range is presumed to stretch for unobservable distances underground. What scientists do know for certain is where you might be able to get a look at them, if you have a practiced eye, the necessary diving gear, and no small amount of good fortune. Living in a single underground aquifer inaccessible to humans, the salamanders are sometimes to be found hiding in rocky pools near the surface, usually after strong rains or inhospitable conditions flush them out of the cavities.

"Flushed" to the surface? This isn't a toilet. And we get where we're going under our own power, thank you very much.

How would you say it then, swimming?

You anthros. You've got it all backwards.

What? Swimming?

No, this thing you call water. You've got it all wrong.

These spring-fed pools in Austin, Texas, are the sole worldwide habitat for these endangered animals.

They are also *literally* a swimming pool; a beloved swimming hole for generations. Bathing at Barton Springs has had added cachet since 1997 when *sosorum*, known as the Barton

Springs salamander, was listed as endangered under the US Endangered Species Act, with the more recently discovered *waterlooensis*, known as the Austin Blind, listed in the summer of 2013.

What do you call it then?

We don't have a word for it.

You're saying your species *doesn't have a word for water?*

You sound irritated.

This wet stuff all around us. The condition of wetness that you're immersed in. It's your home . . . your environment . . . the thing that sustains you. It seems fairly important that you'd have a word for all that, no?

Ah, I think I understand. Yes, we know very well what this thing you call water is but there is no word for it that you would understand. The word for water means universe, lifecurrent, deathcurrent, wavetime, hot, cold, high, low, together, and apart. It is the Sustainer, the Connection, the Fundament.

The message and the medium.

Something like that I suppose.

No, we humans have no word for this. If only we did.

What is this thing you call water?

It depends I guess on who you ask. Fluid, drink, natural resource, solvent, vacation amenity, flood risk—

How strange that sounds.

It isn't the same thing. To everyone, I mean. There are so many different interest groups involved.

But saving these endangered species has proved challenging, as protecting the salamanders means protecting not only the springs themselves but thousands of acres around the city as well, where trace chemicals and contaminants from lawns, vehicles, manufacturing, and even the roads themselves, make their way into the Edwards Aquifer that feeds the pools. It is a controversy with multimillion-dollar implications for residents of local communities, landowners, and developers. Designating this area—one of the fastest-growing urban corridors in the United States—as a critical habitat has profound and lasting repercussions for Austin. It dictates not only who can use Barton Springs and under what conditions, but where and how the city can site new developments and infrastructure, including water treatment facilities, power generation plants, roads, and sewers.

Swimming pools and sewers. So much for an anthro's take on water. What's your species's word for air?

What do you mean?

You know, the airy stuff all around you land creatures. The atmospheric conditions that you're immersed in at all times. It's your oxygenated life substance . . . your climate . . . the thing that sustains you. It seems fairly important that you'd have a word for all that, no?

Point taken.

Even though we Eurycea are aquatic, we have such a concept.

For the climate?

No, not quite. For the condition of swimming in air.

How beautiful.

Not always. For us, it can be the worst death. Prolonged and indescribable, the fatal immersion in a life-giving substance that leads instead to death.

Drowning.

What?

The analogue for us would be drowning.

Or maybe the analogue is the environmental catastrophe you're wreaking on your own life support systems and ours.

Ouch. It's not so easy on the surface, you know. We've got a lot to figure out still about how to manage everything among so many competing needs and interests.

> The case of *Eurycea*, with its complex and historically contingent relationships of people, animals, watersheds, urban developers, federal law, city scientists, and university researchers, illustrates the kind of fluid and nonlinear entanglements to be traced throughout this study. The region's freshwater supply, which in decades past was abundant enough to produce geysers of more than twenty feet high out of newly dug wells, has become an increasing cause for concern after years of severe drought fueled by climate change, the unconventional water and property laws of Texas, and the unique hydrology of the porous limestone of the Edwards Aquifer itself, which is particularly sensitive to changing conditions on the surface.

We're not hopeful.

About what?

That you're going to get it right this time.

What do you mean? You're known as the "salamander that saved a city." An Austin success story that's been emulated in places around the country.

No offense, but we've seen what your "good ideas" have looked like before: foreign missions and pleasure boats, water pumps and ice factories, pesticides and concrete.

I tried to write about that too. The history's not well known though, and there are large gaps in the record of the springs.

> Long before anyone knew of the salamanders' existence, or thought to consider that existence endangered, Barton Springs had been the area's principal attraction. Evidencing layers of human and nonhuman activities and uses, the springs bear witness to various natural resource use priorities and human strategies for management of their environments. Part of a series of artesian spring complexes dotted along the fault lines of the Edwards Aquifer from Salado to near Del Rio, Texas, Barton Springs is thought to have appeared around six million years ago when the Texas Colorado River began cutting into the aquifer below. Archaeological remains found in Zilker Park attest to human presence as early as twelve thousand years ago, though the springs at that time were likely brackish and swampy.

You never came that day.

What?

We were waiting for you.

How did you know that?

Even without ears, we hear stuff you know. The currents carry vibrations of many kinds. We heard them talking.

Those were the biologists.

> By 7,000–6,000 B.C.E., Barton Springs had begun to flow freely, refreshing the flow of Barton Creek into the river two miles away. Signs remain of more extensive human presence around the springs and creek, including tools, shelters, quarries, butchering sites, and rock middens found at excavations

in the park. The accounts of early French and Spanish explorers and settlers describe being guided to the springs by Native Americans along broad, well-trodden paths used by people and wildlife alike. Records from early European settlements in the area suggest that at least twenty-one different tribal groups, particularly the Tonkawa, Lipan Apache, Comanche, and Kiowa, may have relied on the springs, as well as the abundant stone, vegetation, and wildlife found around it.

What happened to you anyway?

I don't really want to get into it.

Why not?

There was an accident.

Oh I'm sorry. We had no idea.

It was a while ago now. Years.

Were you hurt?

I'm getting better. It was about five weeks in the hospital and then some months in a wheelchair. A lot of pool therapy, actually. To get me walking again.

Not long after the arrival of the first colonists to Barton Springs the revolutionary war was declared over and the Republic of Texas born. The obscure frontier town then known as Waterloo, located a couple of miles away and across the river from Barton's tract, was designated the capital of the Texan nation in 1839. The presence of the springs figured into the commissioners' decision to select Waterloo over other more suitable sites for the capital, the selection committee enthusing that the springs were "perhaps the greatest and most convenient water power to be found in the Republic." That same year, the first mill was built when Barton made an agreement permitting Lewis Capt and Company to use the largest spring for a sawmill in exchange for all the lumber Barton or his heirs might want.

What happened?

A window washing truck. I was on a bike.

How awful.

I thought about you a lot while I was there. In the hospital I mean.

You did?

I really did. It was a very nice hospital. New, modern. The windows were these walls of glass that couldn't be opened. I missed fresh air so much, more even than I missed walking. At one point they actually called in a respiratory specialist to see if there was something wrong with my breathing. To pass the hours, I would . . . well, I would talk to you.

After the Civil War, use of the springs intensified again particularly as new manufacturing endeavors were initiated alongside agricultural and livestock uses. Walsh and Brothers acquired the old mill site on the north side of the creek in order to operate a grist mill and rock quarry business, while Michael Paggi's arrival in 1870 led to the damming of the springs for a newfangled activity: the production of artificial ice. In 1881, the Texas Fish Commission created its first fish hatchery at the springs, another Barton Springs venture that was quickly shuttered, this time due to public opposition on nationalist grounds, of stocking German carp in the Texas rivers.

So you never made it.

I never did. I think a day or two later I was able to send a message from the hospital and explain. I don't remember any of it very clearly to be honest.

No, I mean you never made it at all.

To Austin? No. I couldn't go back somehow. Kept making excuses. I just cut the whole thing from my dissertation and moved on to something else.

> Aside from utilitarian uses, the springs had always been an attraction for locals and visitors in the area wishing to fish or swim; the reclusive Barton's baby buffalo were one of the early draws. Throughout the latter part of the nineteenth century, newspaper accounts and memoirs of the time recount the many picnics, parties, and military reunions held at the site. In 1875 the riverboat *Sunbeam* was making regular trips from Austin to ferry pleasure-seekers to Barton Springs, and throughout the 1880s spas and resorts sprung up around the Texan springs, catering to health-conscious vacationers wanting a restorative soak in mineral water. By 1884, complaints of nude bathers at the springs were being voiced—to be repeated a century later when Barton Springs became a hippie hangout.

I came to see that my accident was just one more weird little footnote in the long history of the springs, another crashing of the party, so to speak.

I'm not sure what to say to that.

It's OK. I know you've been here much—much—longer than us.

> From the perspective of *Eurycea*, there is only one fact to be told: that the salamanders have been there continuously since before the people, the cattle, the mills, ice factories, fish hatcheries, mass baptisms, picnics, faux amphitheaters, diving boards, swim meets, the Fish and Wildlife Service,

the mall. The history of the springs is complicated and multilayered, and the perspective one takes has stakes for how one imagines the present and future course of the springs. Depending on one's point of view, either the salamanders have already had to endure too much, their numbers in clear decline since the stories of old when the springs were thick with amphibians, or *Eurycea* are remarkably resistant despite centuries of outside interference. Indeed, the salamanders are thought by evolutionary biologists to predate the springs themselves by millions of years, a relic species reduced to blind living underground when their inland sea home disappeared.

Ah, so you know that part too then.

Yes. It still seems incredible though that your species was here before the springs even existed. Under US law, you basically don't exist now without the springs and vice versa.

What will you do now?

I don't really know. I never wrote the ending.

Our Eurycea *songs all end the same way.*

Surely not "happily ever after"?

Not quite: "This song is sung to the water's edge."

But we don't know where your water's edge is—that's the trouble with trying to protect Barton Springs.

That's why Eurycea *stories always start at the water's edge too.*

Agua de bordes lúbricos

〰〰〰

Coral Bracho

Agua de medusas,
agua láctea, sinuosa,
agua de bordes lúbricos; espesura vidriante —Delicuescencia
entre contornos deleitosos. Agua —agua suntuosa
de involución, de languidez

en densidades plácidas. Agua,
agua sedosa y plúmbea en opacidad, en peso —Mercurial;
      agua en vilo, agua lenta. El alga
acuática de los brillos —En las ubres del gozo. El alga, el
      hálito de su cima;

—sobre el silencio arqueante, sobre los istmos
del basalto; el alga, el hábito de su roce,
su deslizarse. Agua luz, agua pez; el aura, el ágata,
sus desbordes luminosos; Fuego rastreante el alce

huidizo —Entre la ceiba, entre el cardumen; llama
pulsante;
agua lince, agua sargo (El jaspe súbito). Lumbre
entre medusas.
—Orla abierta, labiada; aura de bordes lúbricos,
su lisura acunante, su eflorescerse al anidar; anfibia,
lábil —Agua, agua sedosa
en imantación; en ristre. Agua en vilo, agua lenta —El
      alumbrar lascivo

en lo vadeante oleoso,
sobre los vuelcos de basalto. —Reptar del ópalo entre la
      luz,

entre la llama interna. —Agua
de medusas.
Agua blanda, lustrosa;
agua sin huella; densa,
mercurial
        su blancura acerada, su dilución en alzamientos de grafito,
en despuntar de lisa; hurtante, suave. —Agua viva

su vientre sobre el testuz, volcado sol de bronce envolviendo
—agua blenda, brotante. Agua de medusas, agua táctil
fundiéndose
en lo añil untuoso, en su panal reverberante. Agua amianto, ulva
El bagre en lo mullido
—libando; en el humor nutricio, entre su néctar delicado; el áureo
embalse, el limbo, lo transluce. Agua leve, aura adentro el ámbar
—el luminar ungido, esbelto; el tigre, su pleamar
bajo la sombra vidriada. Agua linde, agua anguila lamiendo
        su perfil,
su transmigrar nocturno
—Entre las sedas matriciales; entre la salvia. —Agua

entre merluzas. Agua grávida (—El calmo goce
tibio; su irisable) —Agua
sus bordes

—Su lisura mutante, su embeleñarse
entre lo núbil
cadencioso. Agua,
agua sedosa de involución, de languidez
en densidades plácidas. Agua, agua;   Su roce
—Agua nutria, agua pez. Agua

de medusas,
agua láctea, sinuosa; Agua,

Water's Lubricious Edges

CORAL BRACHO, TRANSLATION BY FORREST GANDER

Water of jellyfish,
lacteal, sinuous water,
water of lubricious borders; glassy thickness —Deliquescence
in delectable contours. Water —sumptuous water
of involution, of languor

in placid densities. Water,
water silken and plumbeous in opacity, in weight —Mercurial;
    water in suspense, slow water. The algal
bloom brilliant —In the paps of pleasure. The algae, the
    bracing vapor of its peak;

—across the arched silence, across isthmuses
of basalt; the algae, its abrasive rub,
its slippage. Light water, fish water; the aura, the agate,
its luminous border-breakings; Fire trailing the fleeing

elk —Around the ceiba tree, around the shoal of fish; flame
pulsing;
lynx water, sargo water (The sudden jasper.) Luminescence
of jellyfish.
—Edge open, lipped; aura of lubricious borders,
its smoothness rocking, its nesting efflorescence; amphibious,
labile —Water, water silken
with voltaic charge; at rest. Water in suspense, slow water —The
    lascivious luminescence

in its oily breaching,
over faulted basalt. —The slither of opal through the sheen,

through the interior flame. —Water
of jellyfish.
Soft, lustrous water;
traceless water; dense,
mercurial
         its steely whiteness, its dissolution in graphite surges,
in burnished gloss; furtive, smooth. —Living water

upwelling ventral over dorsal, capsized bronze sun enfolding
—crystalline zinc, spouting water. Water of jellyfish, tactile water
fusing itself
to the unctuous indigo blue, to its reverberant honeycomb. Amianthus, ulva water
The catfish in its silt
—sucking; in the nutritious essence, in its delicate nectar; the aureous
reservoir, a limbo, transparentizes it. Light water, aura within amber
—graceful, anointed luminance; the tiger, its high tide
below the brittle shadow. Boundary water, eel water licking
         its profile,
its nocturnal migration
—In silk matrices; in the sage. —Water

between gray-finned hake. Gravid water (—The calm pleasure
tepid; its iridescence) —Water
its borders

—Its shifting smoothness, its enchantment
with what is nubile,
cadenced. Water,
silky water of involution, of languor
in placid densitites.   Water, water;   Its caress
—Otter water, fish water. Water

of jellyfish,
lacteal, sinuous water; Water,

CORAL BRACHO, TRANSLATION BY FORREST GANDER

Contributors

Atul Bhalla is a conceptual artist who lives and works in New Delhi, India. His sculpture, painting, installation, video, photography, and performance works explore the physical, historical, and political significance of water in urban New Delhi. As a Mellon Artist Research Africa Fellow at the University of the Witwatersrand, South Africa, in 2017 and 2018, he examined acid mine drainage and environmental desecration at gold mine sites around Johannesburg. He has exhibited in the USA, Europe, and Asia, and in group exhibitions at the FotoFest Biennale Houston in 2016 and 2108, the Pompidou Center in Paris, the IVAM Institute of Modern Art in Valencia, and the Devi Art Foundation in New Delhi. He has published two books on his performative works: *Yamuna Walk* and *What Will Be My Defeat?*

Coral Bracho is a poet and translator who was born in Mexico City, where she lives and teaches. She is the author of several collections of poetry, including *El ser que va a morir*; *Tierra de entraña ardiente*, a collaboration with painter Irma Palacios; *Ese espacio, ese jardín*, which won the Xavier Villaurrutia Award; and *Cuarto de hotel*. Her *Firefly under the Tongue: Selected Poems* was translated by Forrest Gander. Bracho has been awarded the Aguacalientes National Poetry Prize and a Guggenheim Fellowship.

Akiko Busch is the author of five essay collections, including *Nine Ways to Cross a River*, *The Incidental Steward*, and *How to Disappear: Notes on Invisibility in a Time of Transparency*. She was a contributing editor at *Metropolis* magazine for twenty years, and her essays about design, culture, and nature have appeared in numerous magazines, newspapers, and exhibition catalogs. She has taught at the University of Hartford, Bennington College, and the School of Visual Arts. She lives in the Hudson Valley in New York State.

Colin Channer was born in Jamaica, and was educated there and in New York. He is the author of the poetry collection *Providential* and three books of fiction: *The Girl with the Golden Shoes*, *Passing Through*, and *Waiting in Vain*. Poems from his forthcoming collection of poetry, *Console*, have been published in periodicals including the *New Yorker*, the *Poetry Review*, and *Liberties*. He has led or been part of the planning committees for reading series and festivals in the United States and Senegal; he is also past president of the Jamaican chapter of International PEN. Channer has taught at Pacific University, Brandeis University, and at Wellesley College, and now teaches at Brown University in the Department of Literary Arts.

Ashley Dawson is an author, an activist, and a professor of English at the CUNY Graduate Center and at the College of Staten Island, City University of New York. He specializes in postcolonial studies, cultural studies, and environmental humanities. A long-time climate justice activist, he works to analyze and build alternatives to the current planet-destroying capitalist/colonialist system. His latest books include *People's Power: Reclaiming the Energy Commons*, *Extreme Cities: The Peril and Promise of Urban Life in the Age of Climate Change*, and *Extinction: A Radical History*. He is a member of the Social Text Collective and the founder of the CUNY Climate Action Lab.

Kulvinder Kaur Dhew is the director of Sugar Maples Center for Creative Arts at the Catskill Mountain Foundation in New York. She received an MA (Hons) in Painting from The Royal College of Art during which time she was awarded the RCA Study Grant at Ar.CO in Lisbon, Portugal. A recipient of the prestigious Delfina Fellowship in London, Kulvinder produced large-scale works subsequently collected by MTV Europe, Kazuo Ishiguro, the UK Department of Education and Science, and others. While Head of Painting at The School of Art in Dunedin, New Zealand, she began to take interest in wilderness issues and the then early alarm bells regarding climate change. Later, while living and teaching at Universiti Malaysia Sarawak on the Island of Borneo, she undertook a body of works exploring censorship related to the largest forest fires recorded at the time. Kulvinder continues to explore environmental and cultural themes.

Forrest Gander, born in the Mojave Desert, lives in California. A translator/writer with degrees in geology and literature, he received the Pulitzer Prize, Best Translated Book Award, and fellowships from the Library of Congress, Guggenheim, and US Artists Foundations. His book *Twice Alive* focuses on human and ecological intimacies.

Samuel Gregoire was born in Port-au-Prince, Haiti. Since 2004, he has lived in Santo Domingo, Dominican Republic, where his books are published. He has worked as a human rights journalist and as a writer. His work has appeared in numerous international anthologies, including *Words from an Island* and the *Anthology of Haitian and Dominican Poetry*, edited by Basilio Belliard and Gasthon Saint-Fleur. His poems have also been published in various international literary magazines, including *Poesía, La Libélula vaga, Red Door, Álastor,* and *Abisinia Review.* Gregoire's most recent bilingual book (in Creole and Spanish) is *Simulacros de paraísos y otros renacidos.* Gregoire writes in Spanish, Creole, and French.

Brenda Hillman has written numerous books of poetry. Each of her five most recent collections—*Extra Hidden Life, Among the Days; Seasonal Works with Letters on Fire; Practical Water; Pieces of Air in the Epic;* and *Cascadia*—receives her "sustained attention" to one of the natural elements. In 2016 she was named Academy of American Poets Chancellor. Her awards include the 2012 Academy of American Poets Fellowship, the 2005 William Carlos Williams Prize for poetry, and fellowships from the National Endowment for the Arts and the Guggenheim Foundation. Hillman teaches at Saint Mary's College in Moraga, California, and lives in the Bay Area.

Maya Khosla is a wildlife biologist and writer. Her poetry books include *All the Fires of Wind and Light,* which won the 2020 PEN Oakland/ Josephine Miles Literary Award; *Keel Bone,* which received the Dorothy Brunsman Poetry Prize; and *Heart of the Tearing.* Khosla served as the Poet Laureate of Sonoma County from 2018 through 2020, organizing a series of filmed readings to bring Sonoma's communities together after the 2017 fires. Her poems have been featured in journals, including *River Teeth, California Quarterly,* and *Nomadic Coffee Press,* and have been nominated for Pushcart Prizes and featured in documentary films.

Lenore Manderson is a distinguished professor of public health and medical anthropology in the School of Public Health at the University of the Witwatersrand, Johannesburg, South Africa. From 2014 to 2019 she was also a distinguished visiting professor at Brown University in Providence, Rhode Island, in the Institute at Brown for Environment and Society. She is the author of *Surface Tensions: Surgery, Bodily Boundaries, and the Social Self*, among other books; her most recent coedited work is *Viral Loads: Anthropologies of Urgency in the Time of Covid-19*. She curated "Earth, Itself" at Brown University (2015–2019), and "Watershed" at the University of the Witwatersrand (2018). She is the recipient of the 2023 Bronislaw Malinowski Award of the Society for Applied Anthropology.

Will McGrath's debut, *Everything Lost Is Found Again*, won the Open Borders Book Prize, which celebrates books that explore global ways of thinking and being. His award-winning nonfiction has been translated into several languages, and he has written for *The Atlantic*, *Foreign Affairs*, *Pacific Standard*, *Guernica*, and *Roads & Kingdoms*, among other magazines. He lives with his family in Minneapolis.

José-Luis Moctezuma is a Chicano poet, professor, researcher, and editor. He holds a PhD in English from the University of Chicago. His critical work has been published in *Jacket2*, *Chicago Review*, *Postmodern Culture*, *Modernism/modernity*, and elsewhere. His poetry chapbook, *Spring Tlaloc Seance*, appeared in 2016. His first full-length book of poems, *Place-Discipline*, was the winner of the 2017 Omnidawn 1st/2nd Poetry Book Prize. He lives and teaches in Chicago.

Zoe Nyssa is a poet and an assistant professor of anthropology at Purdue University, where she teaches classes on evidence, expertise, and practice. Her work in collaboration with leading environmental organizations combines traditional ethnographic methods with "big data" techniques to evaluate outcomes for environmental science and policy. Her first book, *A Sense of Emergency and Possibility*, tracks how biodiversity conservation became a science and why this matters. Previously, Nyssa was jointly appointed as Ziff Environmental Fellow at the Harvard University Center for the Environment and the Kennedy School for Government.

Ailsa Piper has worked as a writer, actor, director, broadcaster, teacher, and speaker. She won the Patrick White Playwrights Award for her play *Small Mercies*. Her book *Sinning across Spain* saw her embark on an eight-hundred-mile hike from Granada to Galicia, exploring empathy and communal responsibility by carrying the "sins" of donors in the tradition of medieval pilgrims. Her second book, written with Tony Doherty, a priest, is *The Attachment: Letters from a Most Unlikely Friendship*. Ailsa lives in Sydney, Australia, and swims daily—rain or shine.

Elizabeth Rush is the author of *Rising: Dispatches from the New American Shore*, a finalist for the 2019 Pulitzer Prize for General Nonfiction. Her essays have appeared in *Harper's*, *Granta*, the *New York Times*, *Guernica*, *Creative Nonfiction*, and *Orion*. Rush is the recipient of fellowships from the National Science Foundation, the Andrew Mellon Foundation, the Society for Environmental Journalism, and the Metcalf Institute. In 2016, she was awarded the Howard Foundation Fellowship from Brown University, where she currently teaches courses on writing and reading literary nonfiction.

Cole Swensen is the author of nineteen collections of poetry, most recently *Art in Time*, *On Walking On*, and *Landscapes on a Train*, and a volume of critical essays. Her poetic collections are inspired by research projects on public land use, landscape perception, visual art, and ghosts, among other subjects. A finalist for the National Book Award, her work has won the Iowa Poetry Prize, the San Francisco State Poetry Center Book Award, and a National Poetry Series award, among other prizes. A former Guggenheim Fellow, she is a coeditor of the Norton anthology *American Hybrid* and the founding editor of La Presse Poetry. She has translated twenty titles from French and won the PEN USA Award in Literary Translation. She teaches at Brown University.

Arthur Sze is the author of eleven books of poetry, including *The Glass Constellation: New and Collected Poems* and *Sight Lines*, for which he received the 2019 National Book Award for Poetry. Sze is the recipient of many honors, including the Jackson Poetry Prize, a Lannan Literary Award, a Guggenheim Fellowship, a Lila Wallace–Reader's Digest Writers' Award, two NEA Creative Writing fellowships, and a Howard Foundation Fellowship. A professor emeritus at the Institute of American Indian Arts, he lives in Santa Fe.

Wendy Woodson is a writer, director, choreographer, and video artist, and is the Roger C. Holden Professor of Theater and Dance at Amherst College. Approximately one hundred of her works for stage, videos, and installations have been presented at festivals in the United States, Europe, New Zealand, and Australia; including appearances at the DeCordova Museum in Lincoln, Massachusetts; in Sarajevo, Bosnia; and the Immigration Museum, Victoria, in Melbourne, Australia. She has received fellowships and grants in choreography, playwriting, and video from the National Endowment for the Arts, the Massachusetts Cultural Council, the DC Commission on the Arts and Humanities, the Bellagio Rockefeller Foundation, the Bogliasco Foundation, and the Fulbright Commission.